Animal Breeding

Animal Breeding

by

Rana Vikram Singh

Wordpen Academics

Publishing Science That Matter

ISBN: 978-81-947689-0-6 (Hardback)

Published by : **WORDPEN ACADEMICS**
4736/23, Ansari Road, Darya Ganj,
New Delhi – 110 002
Phone: +91-011-35004336
E-mail: info@wordpenacademics.com

Printed at : Replika Press Pvt. Ltd.

Exclusive Distributor: ***Bio-Green Books, New Delhi***

Preface

In mammals, pregnancy is the period of reproduction during which a female carries one or more live offspring from implantation in the uterus through gestation. It begins when a fertilized zygote implants in the female's uterus; and ends once it leaves the uterus. A male and female copulate, the male inseminating the female. The spermatozoon fertilizes an ovum or various ova in the uterus or fallopian tubes, and this results in one or multiple zygotes. Sometimes, a zygote can be created by humans outside of the animal's body in the artificial process of in-vitro fertilization. After fertilization, the newly formed zygote then begins to divide through mitosis, forming an embryo, which implants in the female's endometrium. At this time, the embryo usually consists of 50 cells. A blastocoele is a small cavity on the center of the embryo, and the developing embryonary cells will grow around it. Then, a flat layer cell forms on the exterior of this cavity, and the zona pellucida, the blastocyst's barrier, remains the same size as before. Cells grow increasingly smaller to fit in. This new structure with a cavity in the center and the developing cells around it is known as a blastocyst.

The presence of the blastocyst means that two types of cells are forming, an inner-cell mass growing on the interior of the blastocele and cells growing on the exterior of it. In 24 to 48 hours, the zona pellucida breaches. The cells on the exterior of the blastocyst begin excreting an enzyme which erodes epithelial uterine lining and creates a site for implantation. The cells surrounding the blastocyst now destroy cells in the uterine lining, forming small pools of blood, which in turn stimulate the production of capillaries. This is the first stage in the growth of the placenta. The inner cell mass of the blastocyst divides rapidly, forming two layers. The top layer becomes the embryo, and cells from there occupy the amniotic cavity. At the same time, the bottom layer forms a small sac (if the cells begin developing in an abnormal position, an ectopic gestation may also occur at this point). Several days later, chorionic villi in the forming placenta anchor the implantation site to the uterus. A system of blood and blood vessels now develops at the point of the newly forming placenta, growing near the implantation site. The small sac inside the blastocyst begins producing red blood cells. For the next 24 hours, connective tissue develops between the developing placenta and the growing embryo. This later develops into the umbilical cord. Following this, a narrow line of cells appears on the surface on the embryo. Its growth makes the embryo undergo gastrulation, in which the three primary tissue layers of the fetus, the ectoderm, mesoderm, and endoderm, develop. The narrow line of cells begin to form the endoderm and mesoderm. The ectoderm begins to grow rapidly as a result of chemicals being produced by the mesoderm. These three layers give rise to all the various types of tissue in the body.

The book is intended to assist students and teachers who are interested in this subject.

— *Dr. Rana Vikram Singh*

Contents

1

Asexual Reproduction

Asexual reproduction is the process by which an organism creates a genetically-similar or identical copy of itself without a contribution of genetic material from another individual. Bacteria divide asexually via binary fission; viruses take control of host cells to produce more viruses; Hydras (invertebrates of the order *Hydroidea*) and yeasts are able to reproduce by budding. These organisms do not have different sexes, and they are capable of "splitting" themselves into two or more individuals. Some 'asexual' species, like hydra and jellyfish, may also reproduce sexually. For instance, most plants are capable of vegetative reproduction—reproduction without seeds or spores—but can also reproduce sexually. Likewise, bacteria may exchange genetic information by conjugation. Other ways of asexual reproduction include parthogenesis, fragmentation and spore formation that involves only mitosis. Parthenogenesis is the growth and development of embryo or seed without fertilization by a male. Parthenogenesis occurs naturally in some species, including lower plants (where it is called apomixis), invertebrates (e.g. water fleas, aphids, some bees and parasitic wasps), and vertebrates (e.g. some reptiles, fish, and, very rarely, birds and sharks). It is sometimes also used to describe reproduction modes in hermaphroditic species which can self-fertilize.

Sexual Reproduction

Sexual reproduction is a biological process by which organisms create descendants that have a combination of genetic material contributed from two (usually) different members of the species. Each of two parent organisms contributes half of the offspring's genetic makeup by creating haploid gametes. Most organisms form two different types of gametes. In these *anisogamous* species, the two sexes are referred to as male (producing sperm or microspores) and female (producing ova or megaspores). In *isogamous species* the gametes are similar or identical in form, but may have separable properties and then may be given other different names. For example, in the green alga, *Chlamydomonas reinhardtii*, there are so-called "plus" and "minus" gametes. A few types of organisms, such as ciliates, have more than two kinds of gametes.

Most animals (including humans) and plants reproduce sexually. Sexually reproducing organisms have two sets of genes for every trait (called alleles). Offspring inherit one allele for each trait from each parent, thereby ensuring that offspring have a combination of the parents' genes. Having two copies of every gene, only one of which is expressed, allows deleterious alleles to be masked, an advantage believed to have led to the evolutionary development of diploidy (Otto and Goldstein).

Allogamy

Allogamy is a term used in the field of biological reproduction describing the fertilization of an ovum from one individual with the spermatozoa of another.

Autogamy

Self-fertilization (also known as autogamy) occurs in hermaphroditic organisms where the two gametes fused in fertilization come from the same individual. They are bound and all the cells merge to form one new gamete.

Mitosis and Meiosis

Mitosis and meiosis are an integral part of cell division. Mitosis occurs in somatic cells, while meiosis occurs in gametes.

Mitosis The resultant number of cells in mitosis is twice the number of original cells. The number of chromosomes in the daughter cells is the same as that of the parent cell.

Meiosis The resultant number of cells is four times the number of original cells. This results in cells with half the number of chromosomes present in the parent cell. A diploid cell duplicates itself, then undergoes two divisions (tetraploid to diploid to haploid), in the process forming four haploid cells. This process occurs in two phases, meiosis I and meiosis II.

Same-sex Reproduction

In recent decades, developmental biologists have been researching and developing techniques to facilitate same-sex reproduction. The obvious approaches, subject to a growing amount of activity, are female sperm and male eggs, with female sperm closer to being a reality for humans, given that Japanese scientists have already created female sperm for chickens. More recently, by altering the function of a few genes involved with imprinting, other Japanese scientists combined two mouse eggs to produce daughter mice.

Reproductive Strategies

There are a wide range of reproductive strategies employed by different species. Some animals, such as the human and Northern Gannet, do not reach sexual maturity for many years after birth and even then produce few offspring. Others reproduce quickly; but, under normal circumstances, most offspring do not survive to adulthood. For example, a rabbit (mature after 8 months) can produce 10–30 offspring per year, and a fruit fly (mature after 10–14 days) can produce up to 900 offspring per year. These two main strategies are known as K-selection (few offspring) and r-selection (many offspring). Which strategy is favoured by evolution depends on a variety of circumstances. Animals with few offspring can devote more resources to the nurturing and protection of each individual offspring, thus reducing the need for many offspring. On the other hand, animals with many offspring may devote fewer resources to each individual offspring; for these types of animals it is common for many offspring to die soon after birth, but enough individuals typically survive to maintain the population.

Other Types of Reproductive Strategies

Polycyclic animals reproduce intermittently throughout their lives.

Semelparous organisms reproduce only once in their lifetime, such as annual plants. Often, they die shortly after reproduction. This is a characteristic of r-strategists. Iteroparous organisms produce offspring in successive (e.g. annual or seasonal) cycles, such as perennial plants. Iteroparous animals survive over multiple seasons (or periodic condition changes). This is a characteristic of K-strategists.

Asexual vs. Sexual Reproduction

Organisms that reproduce through asexual reproduction tend to grow in number exponentially. However, because they rely on mutation for variations in their DNA, all members of the species have similar vulnerabilities. Organisms that reproduce sexually yield a smaller number of offspring, but the large amount of variation in their genes makes them less susceptible to disease.

Many organisms can reproduce sexually as well as asexually. Aphids, slime molds, sea anemones, some species of starfish (by fragmentation), and many plants are examples. When environmental factors are favourable, asexual reproduction is employed to exploit sui conditions for survival such as an abundant food supply, adequate shelter, favorable climate, disease, optimum pH or a proper mix of other lifestyle requirements. Populations of these organisms increase exponentially via asexual reproductive strategies to take full advantage of the rich supply resources.

When food sources have been depleted, the climate becomes hostile, or individual survival is jeopardized by some other adverse change in living conditions, these organisms switch to sexual forms of reproduction. Sexual reproduction ensures a mixing of the gene pool of the species.

The variations found in offspring of sexual reproduction allow some individuals to be better suited for survival and provide a mechanism for selective adaptation to occur. In addition, sexual reproduction usually results in the formation of a life stage that is able to endure the conditions that threaten the offspring of an asexual parent. Thus, seeds, spores, eggs, pupae, cysts or other "over-wintering" stages of sexual reproduction ensure the survival during unfavourable times and the organism can "wait out" adverse situations until a swing back to suitability occurs.

Life Without Reproduction

The existence of life without reproduction is the subject of some speculation. The biological study of how the origin of life led from non-reproducing elements to reproducing organisms is called abiogenesis. Whether or not there were several independent abiogenetic events, biologists believe that the last universal ancestor to all present life on earth lived about 3.5 billion years ago. Today, some scientists have speculated about the possibility of creating life non-reproductively in the laboratory. Several scientists have succeeded in producing simple viruses from entirely non-living materials. The virus is often regarded as not alive. Being nothing more than a bit of RNA or DNA in a protein capsule, they have no metabolism and can only replicate with the assistance of a hijacked cell's metabolic machinery. The production of a truly living organism (*e.g.*, a simple bacterium) with no ancestors would be a much more complex task, but may well be possible according to current biological knowledge.

Lottery Principle

Sexual reproduction has many drawbacks, since it requires far more energy than asexual reproduction and diverts the organisms from other pursuits, and there is some argument about why so many species use it. George C. Williams used lottery tickets as an analogy in one explanation for the widespread use of sexual reproduction. He argued that asexual reproduction, which produces little or no genetic variety in offspring, was like buying many tickets that all have the same number, limiting the chance of "winning"-that is, producing surviving offspring. Sexual reproduction, he argued, was like purchasing fewer tickets but with a greater variety of numbers and therefore a greater chance of success. The point of this analogy is that since asexual reproduction does not produce genetic variations, there is little ability to quickly adapt to a changing environment. The lottery principle is less accepted these days because of evidence that asexual reproduction is more prevalent in unstable environments, the opposite of what it predicts.

Plant Reproduction

Plant reproduction is the production of new individuals or offspring in plants, which can be accomplished by both sexual or asexual means. Sexual reproduction produces offspring by the fusion of gametes, resulting in offspring genetically different from the parent or parents. Asexual reproduction produces new individuals without the fusion of gametes, they are genetically identical to the parent plants and each other, except when mutations occur. In seed plants, the offspring are packaged in a protective seed, which is used as an agent of dispersal.

Asexual Reproduction

Plants have two main types of asexual reproduction in which new plants are produced that are genetically identical clones of the parent individual. "Vegetative" reproduction involves a vegetative piece of the original plant (budding, tillering, etc.) and is distinguished from "apomixis", which is a "replacement" for sexual reproduction, and in some cases involves seeds. Apomixis occurs in many plant species and also in some non-plant organisms. For apomixis and similar processes in non-plant organisms, see parthenogenesis.

Natural vegetative reproduction is mostly a process found in herbaceous and woody perennial plants, and typically involves structural modifications of the stem or roots and in a few species leaves. Most plant species that employ vegetative reproduction, do so as a means to perennialize the plants, allowing them to survive from one season to the next and often facilitating their expansion in size. A plant that persists in a location through vegetative reproduction of individuals constitutes a clonal colony, a single ramet, or apparent individual, of a clonal colony is genetically identical to all others in the same colony. The distance that a plant can move during vegetative reproduction is limited, though some plants can produce ramets from branching rhizomes or stolons that cover a wide area, often in only a few growing seasons. In a sense, this process is not one of "reproduction" but one of survival and expansion of biomass of the individual. When an individual organism increases in size via cell multiplication and remains intact, the process is called "vegetative growth".

However, in vegetative reproduction, the new plants that result are new individuals in almost every respect except genetic. A major disadvantage to vegetative reproduction, is the transmission of pathogens from parent to daughter plants; it is uncommon for pathogens to be transmitted from the plant to its seeds, though there are occasions when it occurs.

Seeds generated by apomixis are a means of asexual reproduction, involving the formation and dispersal of seeds that do not originate from the fertilization of the embryos. Hawkweed (*Hieracium*), dandelion (*Taraxacum*), some Citrus (*Citrus*) and Kentucky blue grass (*Poa pratensis*) all use this form of asexual reproduction. Pseudogamy occurs in some plants that have apomictic seeds, where pollination is often needed to initiate embryo growth, though the pollen contributes no genetic material to the developing offspring. Other forms of apomixis occur in plants also, including the generation of a plantlet in replacement of a seed or the generation of bulbils instead of flowers, where new cloned individuals are produced.

Natural Vegetative Structures

The rhizome is a modified underground stem serving as an organ of vegetative reproduction, e.g. Polypody, Iris, Couch Grass and Nettles.

Prostrate aerial stems, called runners or stolons are important vegetative reproduction organs in some species, such as the strawberry, numerous grasses, and some ferns. Adventitious buds form on roots near the ground surface, on damaged stems (as on the stumps of cut trees), or on old roots. These develop into above-ground stems and leaves. A form of budding called suckering is the reproduction or regeneration of a plant by shoots that arise from an existing root system. Species that characteristically produce suckers include Elm (*Ulmus*), Dandelion (*Taraxacum*), and members of the Rose Family (*Rosa*).

Another type of a vegetative reproduction is the production of bulbs. Plants like onion (*Allium cepa*), hyacinth (*Hyacinth*), narcissus (*Narcissus*) and tulips (*Tulipa*) reproduce by forming bulbs.

Other plants like potatoes (*Solanum tuberosum*) and dahlia (*Dahlia*) reproduce by a method similar to bulbs: they produce tubers.

Gladioli and crocuses (*Crocus*) reproduce by forming a bulb-like structure called a corm.

Human uses of Asexual Reproduction

The most common form of plant reproduction utilized by people is seeds, but a number of asexual methods are utilized which are usually enhancements of natural processes, including: cutting, grafting, budding, layering, division, sectioning of rhizomes or roots, stolons, tillers (suckers) and artificial propagation by laboratory tissue cloning. Asexual methods are most often used to propagate cultivars with individual desirable characteristics that do not come true from seed. Fruit tree propagation is frequently performed by budding or grafting desirable cultivars (clones), onto rootstocks that are also clones, propagated by layering. In horticulture, a "cutting" is a branch that has been cut off from a mother plant below an internode and then rooted, often with the help of a rooting liquid or powder containing hormones. When a full root has formed and leaves begin to sprout anew, the clone is a self-sufficient plant, genetically identical to the mother plant. Examples include cuttings from the stems of blackberries (*Rubus occidentalis*), African violets (*Saintpaulia*), verbenas (*Verbena*) to produce new plants. A related use of cuttings is grafting, where a stem or bud is joined onto a different stem. Nurseries offer for sale trees with grafted stems that can produce four or more varieties of related fruits, including apples. The most common usage of grafting is the propagation of cultivars onto already rooted plants, sometimes the rootstock is used to dwarf the plants or protect them from root damaging pathogens.

Since vegetatively propagated plants are clones, they are important tools in plant research. When a clone is grown in various conditions, differences in growth can be ascribes to environmental effects instead of genetic differences.

Sexual Reproduction

Sexual reproduction involves two fundamental processes, meiosis which rearranges the genes and reduces the number of chromosomes, and fusion of gametes which restores the chromosome to a complete diploid number. In between these two processes, different types of plants vary. In plants and algae that undergo alternation of generations, a gametophyte is the multicellular structure, or phase, that is haploid, containing a single set of chromosomes:

The gametophyte produces male or female gametes (or both), by a process of cell division called mitosis. The fusion of male and female gametes produces a diploid zygote, which develops by repeated mitotic cell divisions into a multicellular sporophyte. Because the sporophyte is the product of the fusion of two haploid gametes, its cells are diploid, containing two sets of chromosomes. The mature sporophyte produces spores by a process called meiosis, sometimes referred to as "reduction division" because the chromosome pairs are separated once again to form single sets. The spores are therefore once again haploid and develop into a haploid gametophyte. In land plants such as ferns, mosses and liverworts the gametophyte is very small, as in ferns and their relatives. In flowering plants (angiosperms) It is reduced to only a few cells, where the female gametophyte (embryo sac) is known as a megagametophyte and the male gametophyte (pollen) is called a micro gametophyte.

History of Sexual Reproduction

Unlike animals, plants are immobile, and cannot seek out sexual partners for reproduction. In the evolution of early plants, abiotic means, including water and wind, transported sperm for reproduction. The first plants were aquatic and released sperm freely into the water to be carried with the currents. Primitive land plants like liverworts and mosses had motile sperm that swam in a thin film of water or were splashed in water droplets from the male reproduction organs onto the female organs. As taller and more complex plants evolved, modifications in the alternation of generations evolved; in the Paleozoic era progymnosperms reproduced by using spores dispersed on the wind. The seed plants including seed ferns, conifers and cordaites, which were all gymnosperms, evolved 350 million years ago; they had pollen grains that contained the male gametes for protection of the sperm during the process of transfer from the male to female parts. It is believed that insects fed on the pollen, and plants thus evolved to use insects to actively carry pollen from one plant to the next. Seed producing plants, which include the angiosperms and the gymnosperms, have heteromorphic alternation of generations with large sporophytes containing much reduced gametophytes. Angiosperms have distinctive reproductive organs called flowers, with carpels, and the female gametophyte is greatly reduced to a female embryo sac, with as few as eight cells. The male gametophyte consists of the pollen grains. The sperm of seed plants are non-motile, except for two older groups of plants, the Cycadophyta and the Ginkgophyta, which have flagellated sperm.

Flowering Plants

Flowering plants are the dominant plant form on land and they reproduce by sexual and asexual means. Often their most distinguishing feature is their reproductive organs, commonly called flowers. Sexual reproduction in flowering plants involves the production of male and female gametes, the transfer of the male gametes to the female ovules in a process called pollination. After pollination occurs, fertilization happens and the ovules grow into seeds with in a fruit. After the seeds are ready for dispersal, the fruit ripens and by various means the seeds are freed from the fruit and after varying amounts of time and under specific conditions the seeds germinate and grow into the next generation. The anther produces male gametophytes, the sperm is produced in pollen grains, which attach to the stigma on top of a carpel, in which the female gametophytes (inside ovules) are located. After the pollen tube grows through the carpel's style, the sex cell nuclei from the pollen grain migrate into the ovule to fertilize the egg cell and endosperm nuclei within the female gametophyte in a process termed double fertilization. The resulting zygote develops into an embryo, while the triploid endosperm (one sperm cell plus two female cells) and female tissues of the ovule give rise to the surrounding tissues in the developing seed. The ovary, which produced the female gametophyte(s), then grows into a fruit, which surrounds the seed(s). Plants may either self-pollinate or cross-pollinate. Nonflowering plants like ferns, moss and liverworts use other means of sexual reproduction.

Adaptations

Flowers of wind pollinated plants tend to lack petals and or sepals. Typically large amounts of pollen are produced and pollination often occurs early in the growing season before leaves can interfere with the dispersal of the pollen. Many trees and all grasses and sedges are wind pollinated, as such they have no need for large fancy flowers. In plants that use insects or other animals to move pollen from one flower to the next, plants have developed greatly modified flower parts to attract pollinators and to facilitate the movement of pollen from one flower to the insect and from the insect back to the next flower. Plants have a number of different means to attract pollinators including colour, scent, heat, nectar glands, eatable pollen and flower shape. Along with modifications involving the above structures two other conditions play a very important role in the sexual reproduction of flowering plants, the first is timing of flowering and the other is the size or number of flowers produced. Often plant species have a few large, very showy flower while others produce many small flowers, often flowers are collected together into large inflorescences to maximize their visual effect, becoming more noticeable to passing pollinators. Flowers are attraction strategies and sexual expressions are functional strategies used to produce the next generation of plants, with pollinators and plants having co-evolved, often to some extraordinary degrees, very often rendering mutual benefit. The largest family of flowering plants is the orchids (Orchidaceae), estimated by some specialists to include up to 35,000 species, which often have highly specialized flowers used to attract insects and facilitate pollination. The stamens are modified to produce pollen in clusters called pollinium, which are attached to insects when crawling into the flower. The flower shapes are modified to force insects to pass by the pollen, which is "glued" to the insect. Some orchids are even more highly specialized, with flower shapes that mimic the shape of insects to attract them to 'mate' with the flowers, a few even have scents that mimic insect pheromones. Another large group of flowering plants is the Asteraceae or sunflower family with close to 22,000 species, which also have highly modified inflorescences that are flowers collected together in heads composed of a composite of individual flowers called florets. Heads with florets of one sex, when the flowers are pistillate or functionally staminate, or made up of all bisexual florets, are called homogamous and can include discoid and liguliflorous type heads. Some radiate heads may be homogamous too. Plants with heads that have florets of two or more sexual forms are called heterogamous and include radiate and disciform head forms, though some radiate heads may be heterogamous too.

Ferns

Ferns typically produce large diploid sporophytes with rhizomes, roots and leaves; and on fertile leaves called sporangium, spores are produced. The spores are released and germinate to produce short, thin gametophytes that are typically heart shaped, small and green in colour. The gametophytes or thallus, produce both motile sperm in the antheridia and egg cells in separate archegonia. After rains or when dew deposits a film of water, the motile sperm are splashed away from the antheridia, which are normally produce on the top side of the thallus, and swim in the film of water to the antheridia where they fertilize the egg. To promote out crossing or cross fertilization the sperm are released before the eggs are receptive of the sperm, making it more likely that the sperm will fertilize the eggs of different thallus. A zygote is formed after fertilization, which grows into a new sporophytic plant. The condition of having separate sporephyte and gametophyte plants is call alternation of generations. Other plants with similar reproductive means include the *Psilotum*, *Lycopodium*, *Selaginella* and *Equisetum*.

Bryophytes

The bryophytes, which include liverworts, hornworts and mosses, reproduce both sexually and vegetatively. The gametophyte is the most commonly known phase of the plant. An early developmental stage in the gametophyte of mosses (immediately following germination of the meiospore) is called the protonema. All are small plants found growing in moist locations and like ferns, have motile sperm with flagella and need water to facilitate sexual reproduction. These plants start as a haploid spore that grows into the dominate form, which is a multicellular haploid body with leaf-like structures that photosynthesize. Haploid gametes are produced in antherida and archegonia by mitosis. The sperm released from the antherida respond to chemicals released by ripe archegonia and swim to them in a film of water and fertilize the egg cells thus producing a zygote. The zygote divides by mitotic division and grows into a sporophyte that is diploid. The multicellular diploid sporophyte produces structures called spore capsules, which are connected by seta to the archegonia. The spore capsules produce spores by meiosis, when ripe the capsules burst open and the spores are released. Bryophytes show considerable variation in their breeding structures and the above is a basic outline. Also in some species each plant is one sex while other species produce both sexes on the same plant.

Sexual Expression

Many plants have evolved a complex sexuality, which is expressed in different combinations of their reproductive organs. Some species have separate male and female individuals, but the majority of plants have both male and female parts in the same flower. Some plants change their gender expression depending on a number of factors like age, time of day, or because of environmental conditions. Plant sexuality also varies within different populations of some species.

Reproductive System

The reproductive system or genital system is a system of organs within an organism which work together for the purpose of reproduction. Many non-living substances such as fluids, hormones, and pheromones are also important accessories to the reproductive system. Unlike most organ systems, the sexes of differentiated species often have significant differences. These differences allow for a combination of genetic material between two individuals, which allows for the possibility of greater genetic fitness of the offspring. The major organs of the human reproductive system include the external genitalia (penis and vulva) as well as a number of internal organs including the gamete producing gonads (testicles and ovaries). Diseases of the human reproductive system are very common and widespread, particularly communicable sexually transmitted diseases.

Most other vertebrate animals have generally similar reproductive systems consisting of gonads, ducts, and openings. However, there is a great diversity of physical adaptations as well as reproductive strategies in every group of vertebrates.

Human Reproductive System

Human reproduction takes place as internal fertilization by sexual intercourse. During this process, the erect penis of the male is inserted into the female's vagina until the male ejaculates semen, which contains sperm, into the female's vagina. The sperm then travels through the vagina and cervix into the uterus or fallopian tubes for fertilization of the ovum. Upon successful fertilization and implantation, gestation of the foetus then occurs within the female's uterus for approximately nine months, this process is known as pregnancy in humans. Gestation ends with birth, the process of birth is known as labour. Labour consists of the muscles of the uterus contracting, the cervix dilating, and the baby passing out the vagina. Human's babies and children are nearly helpless and require high levels of parental care for many years. One important type of parental care is the use of the mammary glands in the female breasts to nurse the baby.

Humans have a high level of sexual differentiation. In addition to differences in nearly every reproductive organ, numerous differences typically occur in secondary sexual characteristics.

Male Reproductive System

Partially shaved erect male genitalia.

1. Testicles,
2. Epididymis,
3. Corpus cavernosa,
4. Foreskin,
5. Frenulum,
6. Urethral opening,
7. Glans penis,
8. Corpus spongiosum,
9. Penis,
10. Scrotum.

The human male reproductive system is a series of organs located outside of the body and around the pelvic region of a male that contribute towards the reproductive process. The primary direct function of the male reproductive system is to provide the male gamete or spermatozoa for fertilization of the ovum. The major reproductive organs of the male can be grouped into three categories. The first category is sperm production and storage.

Production takes place in the testes which are housed in the temperature regulating scrotum, immature sperm then travel to the epididymis for development and storage. The second category are the ejaculatory fluid producing glands which include the seminal vesicles, prostate, and the vas deferens.

The final category are those used for copulation, and deposition of the spermatozoa (sperm) within the female, these include the penis, urethra, vas deferens, and Cowper's gland. Major secondary sexual characteristics includes: larger, more muscular stature, deepened voice, facial and body hair, broad shoulders, and development of an adam's apple. An important sexual hormone of males is androgen, and particularly testosterone.

Female Reproductive System

Photograph of the vulva.

1. Pubic hair (shaved),
2. litoral hood,
3. Clitoris,

4. Labia majora,
5. Labia minora (enclosing the Vaginal Opening),
6. Perineum.

The human female reproductive system is a series of organs primarily located inside of the body and around the pelvic region of a female that contribute towards the reproductive process. The human female reproductive system contains three main parts: the vagina, which acts as the receptacle for the male's sperm, the uterus, which holds the developing fetus, and the ovaries, which produce the female's ova. The breasts are also an important reproductive organ during the parenting stage of reproduction.

The vagina meets the outside at the vulva, which also includes the labia, clitoris and urethra; during intercourse this area is lubricated by mucus secreted by the Bartholin's glands. The vagina is attached to the uterus through the cervix, while the uterus is attached to the ovaries via the fallopian tubes. At certain intervals, typically approximately every 28 days, the ovaries release an ovum, which passes through the fallopian tube into the uterus.

The lining of the uterus, called the endometrium, and unfertilized ova are shed each cycle through a process known as menstruation. Major secondary sexual characteristics include: a smaller stature, a high percentage of body fat, wider hips, development of mammary glands, and enlargement of breasts. Important sexual hormones of females include estrogen and progesterone.

Production of Gametes

The production of gametes takes place within the gonads through a process known as gametogenesis. Gametogenesis occurs when certain types of germ cells undergo meiosis to split the normal diploid number of chromosomes in humans (n=46) into haploids cells containing only 23 chromosomes.

In males this process is known as spermatogenesis and takes place only after puberty in the seminiferous tubules of the testes. The immature spermatozoon or sperm are then sent to the epididymis where they gain a tail and motility. Each of the original diploid germs cells or primary spermatocytes forms four functional gametes which is each capable of fertilization. In females gametogenesis is known as oogenesis which occurs in the ovarian follicles of the ovaries. This process does not produce mature ovum until puberty. In contrast with males, each of the original diploid germ cells or primary oocytes will form only one mature ovum, and three polar bodies which are not capable of fertilization.

It has long been understood that in females, unlike males, all of the primary oocytes ever found in a female will be created prior to birth, and that the final stages of ova production will then not resume until puberty. However, recent scientific data has challenged that hypothesis. This new data indicates that in at least some species of mammal oocytes continue to be replenished in females well after birth.

Development of the Reproductive System

The development of the reproductive system and urinary systems are closely tied in the development of the human fetus. Despite the differences between the adult male and female reproductive system, there are a number of homologous structures shared between them due to their common origins within the fetus. Both organ systems are derived from the intermediate mesoderm. The three main fetal precursors of the reproductive organs are the Wolffian duct, Müllerian ducts, and the gonad. Endocrine hormones are a well known and critical controlling factor in the normal differentiation of the reproductive system. The Wolffian duct forms the epididymis, vas deferns, ductus deferens, ejaculatory duct, and seminal vesicle in the male reproductive system and essentially disappears in the female reproductive system. For the Müllerian Duct this process is reversed as it essentially disappears in the male reproductive system and forms the fallopian tubes, uterus, and vagina in the female system. In both sexes the gonad goes on to form the testes and ovaries, because they are derived from the same undeveloped structure they are considered homologous organs. There are a number of other homologous structures shared between male and female reproductive systems. However, despite the similarity in function of the female fallopian tubes and the male epididymis and vas deferens, they are not homologous but rather analogous structures as they arise from different fetal structures.

Examples of homologous human reproductive organs

Male organ	Female organ	Shared function
Cowper's gland	Bartholin's glands	Lubrication secretions
Penis	Clitoris	Erectile tissue and sensation
Testes	Ovary	Gamete production
Prostate gland	Skene's gland	Ejaculatory fluid and sensation

Diseases of the Human Reproductive System

Like all complex organ systems the human reproductive system is affected by many diseases. There are four main categories of reproductive diseases in humans. They are: 1) genetic or congenital abnormalities, 2) cancers, 3) infections which are often sexually transmitted diseases, and 4) functional problems cause by environmental factors, physical damage, psychological issues, autoimmune disorders, or other causes. The best known type of functional problems include sexual dysfunction and infertility which are both broad terms relating to many disorders with many causes.

Specific reproductive diseases are often symptoms of other diseases and disorders, or have multiple, or unknown causes making them difficult to classify. Examples of unclassifiable disorders include Peyronie's disease in males and endometriosis in females. Many congenial conditions cause reproductive abnormalities but are better known for their other symptoms, these include: Turner syndrome, Klinefelter syndrome, Cystic fibrosis, and Bloom syndrome.

It is also known that disruption of the endocrine system by certain chemical adversely affects the development of the reproductive system and can cause vaginal cancer. Many other reproductive diseases have also been link to exposure to synthetic and environmental chemicals. Common chemicals with known links to reproductive disorders include: lead, dioxin, styrene, toluene, and pesticides.

Examples of Congenital Abnormalities

- Kallmann syndrome-Genetic disorder causing decreased functioning of the sex hormone-producing glands caused by a deficiency of a hormone.
- Cryptorchidism-Absence of one or both testes from the scrotum.
- Androgen insensitivity syndrome-A genetic disorder causing people who are genetically male (i.e. XY chromosome pair) to develop sexually as a female due to an inability to utilize androgen.
- Intersexuality-A person who has genitalia and/or other sexual traits which are not clearly male or female.

Examples of Cancers

- Prostate cancer-Cancer of the prostate gland.
- Breast cancer-Cancer of the mammary gland.
- Ovarian cancer-Cancer of the ovary.

- Penile cancer-Cancer of penis.
- Uterine cancer-Cancer of the uterus.
- Testicular cancer-Cancer of the testicles.
- Cervical Cancer-Cancer of the cervix.

Examples of Infections

- HIV-Infection by the retrovirus known as human immunodeficiency virus.
- Genital warts-Sexually transmitted infection caused by some sub-types of human papillomavirus (HPV).
- Herpes simplex-Sexually transmitted infection caused by a virus called herpes simplex virus (HSV) type 2
- Gonorrhea-Common sexually transmitted disease caused by the Gram-negative bacterium *Neisseria gonorrheae*
- Yeast infection-Infection of the vagina by any species of the fungus genus *Candida.*
- Pelvic inflammatory disease-Painful infection of the female uterus, fallopian tubes, and/or ovaries with associated scar formation and adhesions to nearby tissues and organs.
- Syphilis-Sexually transmitted infection caused by the bacterium *Treponema pallidum.*
- Pubic lice-Infection of the pubic hair by crab lice, *Phthirius pubis.*
- Trichomoniasis-Sexually transmitted infection by the single-celled protozoan parasite *Trichomonas vaginalis.*

Examples of Functional Problems

- Impotence-The inability of a male to produce or maintain an erection.
- Hypogonadism-A lack of function of the gonads, in regards to either hormones or gamete production.
- Ectopic pregnancy-When a fertilized ovum is implanted in any tissue other than the uterine wall.
- Hypoactive sexual desire disorder-A low level of sexual desire and interest.
- Female sexual arousal disorder-A condition of decreased, insufficient, or absent lubrication in females during sexual activity

- Premature ejaculation-A lack of voluntary control over ejaculation.

Other Vertebrates

Vertebrate animals all share key elements of their reproductive systems. They all have gamete producing organs or gonads. These gonads are then connected by oviducts to an opening to the outside of the body, typically the cloaca, but sometime to a unique pore such as a vagina or intromittent organ.

Mammals

Most mammal reproductive systems are similar, however, there are some notable differences between the "normal" mammal and humans. For instance, most mammalian males have a penis which is stored internally until erect, and most have a penis bone or baculum. Additionally, males of most species do not remain continually sexually fertile as humans do. Like humans, most groups of mammals have descended testicles found within a scrotum, however, others have descended testicles that rest on the ventral body wall, and a few groups of mammals, such as elephants, have undescended testicles found deep within their body cavities near their kidneys.

Marsupials are unique in that the female has two vaginae, both of which open externally through one orifice but lead to different compartments within the uterus; males usually have a two-pronged penis which corresponds to the females' two vaginae. Marsupials typically develop their offspring in an external pouch containing teats to which their newborn young (joeys) attach themselves for post uterine development. Also, marsupials have a unique prepenial scrotum. The uterus and vagina are unique to mammals with no homologue in birds, reptiles, amphibians, or fish. In place of the uterus the other vertebrate groups have an unmodified oviduct leading directly to a cloaca, which is a shared exit-hole for gametes, urine, and feces. Monotremes (i.e. platypus and echidnas), a group of egg-laying mammals, also lack a uterus and vagina, and in that respect have a reproductive system resembling that of a reptile.

Birds

Male and female birds have a cloaca, an opening through which eggs, sperm, and wastes pass. Intercourse is performed by pressing the lips of the cloacae together, which is sometimes known as the "cloacal kiss", during which time the male transfers his sperm to the female. A few species of birds (e.g. most waterfowl) have a intromittent organ which is known as a phallus that is analogous to the mammals' penis. The female lays amniotic eggs in which the young gestate. Unlike most vertebrates female birds typically have only one functional ovary and oviduct. As a group, birds, like mammals, are noted for their high level of parental care.

Reptiles

Reptiles are almost all sexually dimorphic, and exhibit internal fertilization through the cloaca. Some reptiles lay eggs while others are viviparous (animals that deliver live young). Reproductive organs are found within the cloaca of reptiles. Most male reptiles have copulatory organs, which are usually retracted or inverted and stored inside the body. In turtles and crocodilians, the male has a single median penis-like organ, while male snakes and lizards each possess a pair of penis-like organs.

Amphibians

Most amphibians exhibit external fertilization of eggs, typically within the water, though some amphibians such as caecilians have internal fertilization. All have paired, internal gonads, connected by ducts to the cloaca.

Fish

Fish exhibit a wide range of different reproductive strategies. Most fish however are oviparous and exhibit external fertilization. In this process, females use their cloaca to release a large quantities their gametes, called spawn, into the water and one or more males release "milt", a white fluid containing many sperm over the unfertilized eggs. Other species of fish are oviparous and have internal fertilization aided by pelvic or anal fins that are modified into an intromittent organ analogous to the human penis. A small portion of fish species are either viviparous or ovoviviparous, and are collectively known as livebearers.

Fish gonads are typically pairs of either ovaries or testes. Most fish are sexually dimorphic but some species are hermaphroditic or unisexual.

Invertebrates

Invertebrates have an extremely diverse array of reproductive systems, the only commonality may be that they all lay eggs. Also, aside from cephalopods, and arthropods, nearly all other invertebrates are hermaphroditic and exhibit external fertilization.

Cephalopods

All cephalopods are sexually dimorphic and reproduce by laying eggs. Most cephalopods have semi-internal fertilization in which the male places his gametes inside the female's mantle cavity or pallial cavity to fertilize the ova found in the female's single ovary. Likewise, male cephalopods have only a single teste. In the female of most cephalopods the nidamental glands aid in development of the egg.

The "penis" in most unshelled male cephalopods (Coleoidea) is a long and muscular end of the gonoduct used to transfer spermatophores to a modified arm called a hectocotylus. That in turn is used to transfer the spermatophores to the female. In species where the hectocotylus is missing, the "penis" is long and able to extend beyond the mantle cavity and transfers the spermatophores directly to the female.

Many cephalopods shed their gonads during reproduction, and thus only reproduce once. Most cephalopods die after reproducing. Females nautilus however, have the ability to regenerate their gonads, making them the only cephalopods to spawn once per year. The females in many cephalopod species exhibit some level of parental protection for their eggs.

Prototheria

Prototheria is a taxonomic group, or taxon, to which the order Monotremata belongs. It is conventionally ranked as a subclass within the mammals.

Most of the animals in this group are extinct. The egg-laying monotremes are known from fossils of the Cretaceous and Cenozoic periods; they are represented today by the platypus and several species of echidna.

The names Prototheria, Metatheria and Eutheria (meaning "first beasts", "changed beasts", and "true beasts") refer to the three mammalian groupings which have living representatives. Each of the three may be defined as a putative clade comprising a living crown-group (respectively the Monotremata, Marsupialia and Placentalia) plus any fossil species which are more closely related to that crown-group than to any other animals.

The threefold division of living mammals into monotremes, marsupials and placentals was already well established when Thomas Huxley proposed the names Metatheria and Eutheria to incorporate the two latter groups in 1880. Initially treated as subclasses, Metatheria and Eutheria are by convention now grouped as infraclasses of the subclass Theria, and in more recent proposals have been demoted further (to cohorts or even magnorders), as cladistic reappraisals of the relationships between living and fossil mammals have suggested that the Theria itself should be reduced in rank.

Prototheria, on the other hand, was generally recognised as a subclass until quite recently, on the basis of an hypothesis which defined the group by two supposed synapomorphies: (1) formation of the side wall of the braincase from a bone called the anterior lamina, contrasting with the alisphenoid in therians; and (2) a linear alignment of molar cusps, contrasting with a triangular arrangement in therians. These characters appeared to unite monotremes with a range of Mesozoic fossil orders (Morganucodonta, Triconodonta, Docodonta and Multituberculata) in a broader clade for which the name Prototheria was retained, and of which monotremes were thought to be only the last surviving branch.

The evidence which was held to support this grouping is now universally discounted. In the first place, examination of embryos has revealed that the development of the braincase wall is essentially identical in therians and in 'prototherians': the anterior lamina simply fuses with the alisphenoid in therians, and therefore the 'prototherian' condition of the braincase wall is primitive for all mammals while the therian condition can be derived from it.

Additionally, the linear alignment of molar cusps is also primitive for all mammals. Therefore, neither of these states can supply a uniquely shared derived character which would support a 'prototherian' grouping of orders in contradistinction to Theria.

In a further reappraisal, the molars of embryonic and fossil monotremes (living monotreme adults are toothless) appear to demonstrate an ancestral pattern of cusps which is similar to the triangular arrangement observed in therians. Some peculiarities of this dentition support an alternative grouping of monotremes with certain recently-discovered fossil forms into a proposed new clade known as the Australosphenida, and also suggest that the triangular array of cusps may have evolved independently in australosphenidans and therians.

The Australosphenida hypothesis remains controversial, and some taxonomists prefer to maintain the name Prototheria as a fitting contrast to the other group of living mammals, the Theria. In theory, the Prototheria is taxonomically redundant, since Monotremata is currently the only order which can still be confidently included, but its retention might be justified if new fossil evidence, or a re-examination of known fossils, enables extinct relatives of the monotremes to be identified and placed within a wider grouping.

Theria

Theria is a subclass of mammals that give birth to live young without using a shelled egg, including both eutherians (placental mammals) and metatherians (marsupials and their ancestors).

Extent

Members of this subclass have external ears, most can suckle on a nipple, and have an ankle specialized for power and range of motion. Therians are often classified by their specialized dentition. Almost all currently extant (not extinct) mammals are therians. The only known exceptions are the platypus and the echidnas (spiny anteater), both of which are prototherian monotremes.

Taxonomy

The rank of 'Theria' may vary depending on the classification system used. The textbook classification system by Vaughan et al. (2000) gives the following:

Class Mammalia

- Subclass Theria: live-bearing mammals;
  - o Infraclass Metatheria: marsupials
  - o Infraclass Eutheria: placentals.

In the above system Theria is a subclass. Alternatively, in the system proposed by McKenna and Bell (1997) it is ranked as a supercohort under the subclass Theriiformes:

Class Mammalia

- Subclass Theriiformes: live-bearing mammals and their prehistoric relatives;
  - o Infraclass Holotheria: modern live-bearing mammals and their prehistoric relatives.
    - * Legion Cladotheria
    - * Sublegion Zatheria
    - * Infralegion Tribosphenida
    - * Supercohort Theria: therian mammals

* Cohort Marsupialia: marsupials
* Cohort Placentalia: placentals.

Another classification proposed by Luo et al. (2002) does not assign any rank to the taxonomic levels, but uses a purely cladistic system instead.

Marsupial

Marsupials are an infraclass of mammals, characterized by a distinctive pouch (called the *marsupium*), in which females carry their young through early infancy.

History

It was once commonly believed that marsupials were a primitive forerunner of modern placental mammals, but fossil evidence, first presented by researcher M.J. Spechtt in 1982, conflicts with this assumption. Instead, both main branches of the mammal tree appear to have evolved concurrently toward the end of the Mesozoic era. In the absence of soft tissues, such as the pouch and reproductive system, fossil marsupials can be distinguished from placentals by the form of their teeth; primitive marsupials possess four pairs of molar teeth in each jaw, whereas placental mammals never have more than three pairs. Using this criterion, the earliest known marsupial is *Sinodelphys szalayi*, which lived in China around 125 million years ago. This makes it almost contemporary to the earliest placental fossils, which have been found in the same area.

The discovery of Chinese marsupials appears to support the idea that marsupials reached Australia via Southeast Asia. There are a few species of marsupials still living in Asia, especially in the Sulawesi region of Indonesia. These marsupials coexist with primates, hooved mammals and other placentals. However, due to the fact that Australia and China were separated by the wide Tethys Sea in the early Cretaceous into the Northern continent of Laurasia and Southern continent of Gondwana, marsupials had to take a much longer route around. From their origin in East Laurasia (modern day China), they spread westwards into modern North America (still attached to Eurasia) and skipped across to South America, which was connected to North America up until around 65MYA. Here they radiated into Borhyaenids and Shrew Opossums, creating a unique fauna found in South America and Antarctica (which were connected until 35MYA). Marsupials reached Australia via Antarctica about 50MYA just after Australia had split off, suggesting a single dispersion event of several of just one species, related to South America's Monito del Monte (Microbiothere), rafted across the widening, but still narrow gap between Australia and Antarctica at that time. In Australia, being the only mammals present (except a few Austrosphenids like echidnas and platypuses) they radiated into the wide varieties we see today, even island hopping some way through the Indonesian archipelagos, almost completing a circumnavigation back to their homeland in China.

On most continents, placental mammals were much more successful and no marsupials survived, though in South America the opossums retained a strong presence, and the Tertiary saw the genesis of marsupial predators such as the borhyaenids and the saber-toothed *Thylacosmilus*. In Australia, however, marsupials displaced placental mammals entirely, and have since dominated the Australian ecosystem. Marsupial success over placental mammals in Australia has been attributed to their comparatively low metabolic rate, a trait which would prove helpful in the hot Australian climate. As a result, native Australian placental mammals (such as hopping mice) are more recent immigrants.

Description

An early birth removes a developing marsupial from its parent's body much sooner than in placental mammals, and thus marsupials have not developed a complex placenta to protect the embryo from its mother's immune system. Though early birth places the tiny newborn marsupial at a greater environmental risk, it significantly reduces the dangers associated with long pregnancies, as there is no need to carry a large fetus to full-term in bad seasons.

Because newborn marsupials must climb up to their mother's nipples, their front limbs are much more developed than the rest of the body at the time of birth. It is possible that this requirement has resulted in the limited range of locomotor adaptations in marsupials compared to placentals. Marsupials must develop a grasping forepaw during their early youth, making the transition from this limb into a hoof, wing, or flipper, as some groups of placental mammals have done, far more difficult.

There are about 334 species of marsupial, and over 200 are native to Australia and neighbouring northern islands. There are also 100 extant American species; these are centred mostly in South America, but the Great American Interchange has provided Central America with 13 species, and North America with one (the Virginia Opossum).

Some structural features can be found in marsupials. Ossified patellae are absent. Marsupials (and also monotremes) also lack a gross communication (corpus callosum) between the right and left brain hemisphere.

Reproductive System

Marsupials' reproductive systems differ markedly from those of placental mammals (*Placentalia*). Females have two lateral vaginas, which lead to separate uteruses, but both open externally through the same orifice. A third canal, the median vagina, is used for birth. This canal can be transitory or permanent. The males generally have a two-pronged penis, which corresponds to the females' two vaginas. The penis is used only for discharging semen into females, and there is instead a urogenital sac used to store waste before expulsion.

Pregnant females develop a kind of yolk sac in their wombs, which delivers nutrients to the embryo. Marsupials give birth at a very early stage of development (about 4–5 weeks); after birth, newborn marsupials crawl up the bodies of their mothers and attach themselves to a nipple, which is located on the underside of the mother either inside a pouch called the *marsupium* or open to the environment. To crawl to the nipple and attach to it the marsupial must have well developed forelimbs and facial structures. This is accomplished by accelerating forelimb and facial development in marsupials compared to placental mammals. A result there is decelerated development of such structures as the hindlimb and brain. There they remain for a number of weeks, attached to the nipple. The offspring are eventually able to leave the *marsupium* for short periods, returning to it for warmth, protection and nourishment.

Taxonomy

Taxonomically, there are two primary divisions of Marsupialia: American marsupials and the Australian marsupials. The Order Microbiotheria (which has only one species, the Monito del Monte) is found in South America but is believed to be more closely related to the Australian marsupials. There are many small arboreal species in each group.

The term *opossums* is properly used to refer to the American species (though *possum* is a common diminutive), while similar Australian species are properly called *possums*.

The Virginia Opossum (*Didelphis virginiana*), the only North American marsupial north of Mexico.

- Superorder Ameridelphia;
  - o Order Didelphimorphia (93 species)
    - * Family Didelphidae: opossums.
  - o Order Paucituberculata (6 species)
    - * Family Caenolestidae: shrew opossums.
- Superorder Australidelphia
  - o Order †Yalkaparidontia
  - o Order Microbiotheria (1 species)
    - * Family Microbiotheriidae: Monito del Monte.
  - o Order Dasyuromorphia (71 species)
    - * Family †Thylacinidae: Thylacine
    - * Family Dasyuridae: antechinuses, quolls, dunnarts, Tasmanian Devil, and relatives
    - * Family Myrmecobiidae: Numbat.
  - o Order Peramelemorphia (24 species)
    - * Family Thylacomyidae: bilbies
    - * Family †Chaeropodidae: Pig-footed Bandicoot
    - * Family Peramelidae: bandicoots and allies.
  - o Order Notoryctemorphia (2 species)
    - * Family Notoryctidae: marsupial moles .
  - o Order Diprotodontia (137 species)
    - * Family Phascolarctidae: Koala
    - * Family Vombatidae: wombats
    - * Family †Diprotodontidae: diprotodon
    - * Family Phalangeridae: brushtail possums and cuscuses
    - * Family Burramyidae: pygmy possums
    - * Family Tarsipedidae: Honey Possum
    - * Family Petauridae: Striped Possum, Leadbeater's Possum, Yellow-bellied Glider, Sugar Glider, Mahogany Glider, Squirrel Glider
    - * Family Pseudocheiridae: ringtailed possums and relatives
    - * Family Potoridae: potoroos, rat kangaroos, bettongs

  * Family Acrobatidae: Feathertail Glider and Feather-tailed Possum
  * Family Hypsiprymnodontidae: Musky Rat-kangaroo
  * Family Macropodidae: kangaroos, wallabies, and relatives
  * Family †Thylacoleonidae: marsupial lions.
- o Order †Sparassodonta.

Mammaliaformes

Mammaliaformes ("mammal-shaped") is a clade that contains the mammals and their closest extinct relatives. Phylogenetically it is defined as a clade including the most recent common ancestor of *Sinoconodon*, morganuconodonts, docodonts, Monotremata, Marsupialia, Placentalia, extinct members of this clade, and all of its descendants. The precise phylogeny is disputed due to the scantness of evidence in the fossil record. However, it is thought that the Mammaliaformes were of three major groups: Allotheria, the longest extinct lineage of pre-mammals; Docodonta, including close relatives such as Morganucodonta; and Symmetrodonta, the most basal of modern mammals. Mammaliaformes radiated from Cynodontia. The Probainognathians of the Eucynodont clade probably evolved into the early mammaliaformes, but the branch Allotheria was so different that they may have come from an entirely different group of cynodonts.

Early mammaliforms were generally shrew-like in appearance and size, and most of their distinguishing characteristics were internal. In particular, the structure of the mammaliform (and mammal) jaw and arrangement of teeth is nearly unique. Instead of having many teeth that are frequently replaced, mammals have one set of baby teeth and later one set of adult teeth which fit together precisely. This is thought to aid in the grinding of food to make it quicker to digest. Being warm-blooded requires more calories than "cold-blooded" animals, so quickening the pace of digestion is a necessity. The drawback to the fixed dentition is that worn teeth cannot be replaced, as was possible for the reptilian ancestors of mammaliforms. However, small mammals generally being very short-lived compared to reptiles of the same size, this was not much of a problem during the early phase of their evolution, in which the trait was set. Early mammaliaformes were probably nocturnal.

Mammaliforms have several common structures. Most importantly, mammaliforms have highly specialized molars, with cusps and flat regions for grinding food. This system is also unique to mammals, although it seems to have evolved convergently in pre-mammals multiple times.

Lactation and fur, along with other characteristically mammalian features, are also thought to characterize the Mammaliaformes, but these traits are difficult to study in the fossil record. The fossilized remains of *Castorocauda lutrasimilis* are a unique exception.

Some non-mammal mammaliformes still retain reptile-like traits. Some mammaliformes had reptile-like locomotion. Furthermore, these mammaliformes still had some bones on their lower jaw seen in reptiles.

Monotreme

Monotremes are mammals that lay eggs (Prototheria) instead of giving birth to live young like marsupials (Metatheria) and placental mammals (Eutheria).

They are conventionally treated as comprising a single order Monotremata, though a recent classification proposes to divide them into the orders Platypoda (the Platypus along with its fossil relatives) and Tachyglossa (the echidnas, or spiny anteaters). The entire grouping is also traditionally placed into a subclass Prototheria, which was extended to include several fossil orders but these are no longer seen as constituting a natural group allied to monotreme ancestry. A controversial hypothesis now relates the monotremes to a different assemblage of fossil mammals in a clade termed Australosphenida.

General Characteristics

Like other mammals, monotremes are warm-blooded with a high metabolic rate (though not as high as other mammals, see below); have hair on their bodies; produce milk through mammary glands to feed their young; have a single bone in their lower jaw; and have three middle-ear bones. Monotremes were very poorly understood for many years, and to this day some of the 19th century myths that grew up around them endure. It is still sometimes thought, for example, that the monotremes are "inferior" or quasi-reptilian, and that they are a distant ancestor of the "superior" placental mammals. It now seems clear that modern monotremes are the survivors of an early branching of the mammal tree; a later branching is thought to have led to the marsupial and placental groups.

In common with reptiles and marsupials, monotremes lack the connective structure (corpus callosum) which allows communication between the right and left brain hemispheres in placentals. The key anatomical difference between monotremes and other mammals is the one that gave them their name; *monotreme* means 'single opening' in Greek and comes from the fact that their urinary, defecatory, and reproductive systems all open into a single duct, the cloaca. This structure is very similar to the one found in reptiles. Monotremes and marsupials have a single cloaca (though marsupials also have a separate genital tract), while placental mammal females have separate openings for reproduction, urination, and defecation: the vagina, the urethra, and the anus.

Monotremes lay eggs. However, the egg is retained for some time within the mother, who actively provides the egg with nutrients. Monotremes also lactate, but have no defined nipples, excreting the milk from their mammary glands via openings in their skin. All species are long-lived, with low rates of reproduction and relatively prolonged parental care of infants. Infant echidnas are sometimes known as *puggles*, referencing their similarity in appearance to the Australian children's toy designed by Tony Barber. The same term, though not generally accepted, is popularly applied to young platypuses as well.

Extant monotremes lack teeth as adults. Fossil forms and modern platypus young have "tribosphenic" molars (with the occlusal surface formed by three cusps arranged in a triangle), which are one of the hallmarks of extant mammals. Some recent work suggests that monotremes acquired this form of molar *independently* of placental mammals and marsupials, although this is not well established. The jaw of monotremes is constructed somewhat differently from that of other mammals, and the jaw opening muscle is different. As in all true mammals, the tiny bones that conduct sound to the inner ear are fully incorporated into the skull, rather than lying in the jaw as in cynodonts and other pre-mammalian synapsids; this feature, too, is now claimed to have evolved independently in monotremes and therians, although, as with the analogous evolution of the tribosphenic molar, this is disputed. The external opening of the ear still lies at the base of the jaw. The imminent sequencing of the Platypus genome should shed light on this and many other questions regarding the evolutionary history of the monotremes. The monotremes also have extra bones in the shoulder girdle, including an interclavicle and coracoid, which are not found in other mammals. Monotremes retain a reptile-like gait, with legs that are on the sides of rather than underneath the body. The monotreme leg bears a spur in the ankle region; the spur is nonfunctional in echidnas, but contains a powerful venom in the male Platypus.

Physiology

It is still sometimes said that monotremes have less developed internal temperature control mechanisms than other mammals, but recent research shows that monotremes maintain a constant body temperature in a wide variety of circumstances without difficulty (for example, the Platypus while living in an icy mountain stream). Early researchers were misled by two factors: firstly, monotremes maintain a lower average temperature than most mammals (around 32 °C (90 °F), compared to about 35 °C (95 °F) for marsupials, and 37 °C (99 °F) for most placentals); secondly, the Short-beaked Echidna (which is much easier to study than the reclusive Platypus) maintains normal temperature only when it is active: during cold weather, it conserves energy by "switching off" its temperature regulation. Finally, poor thermal regulation has also been observed in the hyraxes, which are placental mammals. Their metabolic rate is remarkably low by mammalian standards. The Platypus has an average body temperature of about 32 °C (90 °F) rather than the 37 °C (99 °F) typical of placental mammals. Research suggests this has been a gradual adaptation to harsh environmental conditions on the part of the small number of surviving monotreme species rather than a historical characteristic of monotremes. Contrary to previous research, the Echidna does enter REM sleep, albeit only when the ambient temperature of its environment is around 25 °C (77 °F). At the temperatures of 15 °C (59 °F) and 28 °C (82 °F), REMS is suppressed.

Taxonomy

The only surviving examples of monotremes are all indigenous to Australia and New Guinea, although there is evidence that they were once more widespread. Fossil and genetic evidence shows that the monotreme line diverged from other mammalian lines about 150 million years ago and that both the short-beaked and long-beaked echidna species are derived from a platypus-like ancestor. Fossils of a jaw fragment 110 million years old were found at Lightning Ridge, New South Wales.

These fragments, from species *Steropodon galmani*, are the oldest known fossils of monotremes. Fossils from the genera *Kollikodon*, *Teinolophos*, and *Obdurodon* have also been discovered. In 1991, a fossil tooth of a 61-million-year-old platypus was found in southern Argentina (since named *Monotrematum*, though it is now considered to be an *Obdurodon* species). Molecular clock and fossil dating suggest echidnas split from platypuses 19–48 million years ago.

- Order Monotremata;
  - o Suborder Platypoda.
    - * Family Ornithorhynchidae: platypus
    - * Genus *Ornithorhynchus*
    - * Platypus, *Ornithorhynchus anatinus.*
  - o Suborder Tachyglossa.
    - * Family Tachyglossidae: echidnas
    - * Genus *Tachyglossus*
    - * Short-beaked Echidna, *Tachyglossus aculeatus*
    - * *Tachyglossus aculeatus aculeatus*
    - * *Tachyglossus aculeatus acanthion*
    - * *Tachyglossus aculeatus lawesii*
    - * *Tachyglossus aculeatus multiaculeatus*
    - * *Tachyglossus aculeatus setosus*
    - * Genus *Zaglossus*
    - * Sir David's Long-beaked Echidna, *Zaglossus attenboroughi*
    - * Eastern Long-beaked Echidna, *Zaglossus bartoni*
    - * *Zaglossus bartoni bartoni*
    - * *Zaglossus bartoni clunius*
    - * *Zaglossus bartoni diamondi*
    - * *Zaglossus bartoni smeenki*
    - * Western Long-beaked Echidna, *Zaglossus bruijni.*

Fossil Monotremes

The fossil record of monotremes is relatively sparse. Although biochemical and anatomical evidence suggests that monotremes diverged from the mammalian lineage before the marsupials and placental mammals arose, only a handful of monotreme fossils are known from before the Miocene epoch. The few Mesozoic fossils that do exist, such as that of *Steropodon*, seem to indicate that the monotremes first evolved in Australia, during the Late Jurassic or Early Cretaceous. They subsequently spread to both South America and Antarctica, which were still united with Australia at that time, but may not have survived on either continent for long.

2

Pregnancy in Mammals

In mammals, pregnancy is the period of reproduction during which a female carries one or more live offspring from implantation in the uterus through gestation. It begins when a fertilized zygote implants in the female's uterus; and ends once it leaves the uterus.

Fertilization and Implantation

A male and female copulate, the male inseminating the female. The spermatozoon fertilizes an ovum or various ova in the uterus or fallopian tubes, and this results in one or multiple zygotes. Sometimes, a zygote can be created by humans outside of the animal's body in the artificial process of in-vitro fertilization. After fertilization, the newly formed zygote then begins to divide through mitosis, forming an embryo, which implants in the female's endometrium. At this time, the embryo usually consists of 50 cells.

Development

After Implantation

A blastocoele is a small cavity on the center of the embryo, and the developing embryonary cells will grow around it. Then, a flat layer cell forms on the exterior of this cavity, and the zona pellucida, the blastocyst's barrier, remains the same size as before. Cells grow increasingly smaller to fit in. This new structure with a cavity in the center and the developing cells around it is known as a blastocyst.

The presence of the blastocyst means that two types of cells are forming, an inner-cell mass growing on the interior of the blastocele and cells growing on the exterior of it. In 24 to 48 hours, the zona pellucida breaches. The cells on the exterior of the blastocyst begin excreting an enzyme which erodes epithelial uterine lining and creates a site for implantation.

Placental Circulation System

The cells surrounding the blastocyst now destroy cells in the uterine lining, forming small pools of blood, which in turn stimulate the production of capillaries. This is the first stage in the growth of the placenta. The inner cell mass of the blastocyst divides rapidly, forming two layers.

The top layer becomes the embryo, and cells from there occupy the amniotic cavity. At the same time, the bottom layer forms a small sac (if the cells begin developing in an abnormal position, an ectopic gestation may also occur at this point).

Several days later, chorionic villi in the forming placenta anchor the implantation site to the uterus. A system of blood and blood vessels now develops at the point of the newly forming placenta, growing near the implantation site. The small sac inside the blastocyst begins producing red blood cells. For the next 24 hours, connective tissue develops between the developing placenta and the growing embryo. This later develops into the umbilical cord.

Cellular Differentiation

Following this, a narrow line of cells appears on the surface on the embryo. Its growth makes the embryo undergo gastrulation, in which the three primary tissue layers of the fetus, the ectoderm, mesoderm, and endoderm, develop. The narrow line of cells begin to form the endoderm and mesoderm. The ectoderm begins to grow rapidly as a result of chemicals being produced by the mesoderm. These three layers give rise to all the various types of tissue in the body.

The endoderm later forms the lining of the tongue, digestive tract, lungs, bladder and several glands. The mesoderm forms muscle, bone, and lymph tissue, as well as the interior of the lungs, heart, and reproductive and excretory systems. It also gives rise to the spleen, and produces blood cells. The ectoderm forms the skin, nails, hair, cornea, lining of the internal and external ear, nose, sinuses, mouth, anus, teeth, pituitary gland, mammary glands, eyes, and all parts of the nervous system.

Approximately 18 days after fertilization, the embryo has divided to form much of the tissue it will need. It is shaped like a pear, where the head region is larger than the tail. The embryo's nervous system is one of the first organic systems to grow. It begins growing in a concave area known as the neural groove.

The blood system continues to grow networks which allow the blood to flow around the embryo. Blood cells are already being produced and are flowing through these developing networks. Secondary blood vessels also begin to develop around the placenta, to supply it with more nutrients. Blood cells begin to form on the sac in the center of the embryo, as well as cells which begin to differentiate into blood vessels. Endocardial cells begin to form the myocardium.

At about 24 days past fertilization, there is a primitive S-shaped tubule heart which begins beating. The flow of fluids throughout the embryo begins at this stage.

Mammals

There are three extant kinds of mammals: Monotremes, Placentals and Marsupials, all with internal fertilisation. In placental mammals, offspring are born as juveniles: complete animals with the sex organs present although not reproductively functional. After several months or years, the sex organs develop further to maturity and the animal becomes sexually mature. Most female mammals are only fertile during certain periods during their estrous cycle, at which point they are ready to mate. Individual male and female mammals meet and carry out copulation. For most mammals, males and females exchange sexual partners throughout their adult lives.

Male

The male reproductive system contains two main divisions: the penis, and the testicles, the latter of which is where sperm are produced. In humans, both of these organs are outside the abdominal cavity, but they can be primarily housed within the abdomen in other animals (for instance, in dogs, the penis is internal except when mating). Having the testicles outside the abdomen best facilitates temperature regulation of the sperm, which require specific temperatures to survive. Sperm are the smaller of the two gametes and are generally very short-lived, requiring males to produce them continuously from the time of sexual maturity until death. Prior to ejaculation the produced sperm are stored in the seminal vesicle, a small gland that is located just behind the bladder.

A sperm cell is motile and swims via chemotaxis, using its flagellum to propel itself towards the ovum.

Female

The female reproductive system likewise contains two main divisions: the vagina and uterus, which act as the receptacle for the sperm, and the ovaries, which produce the female's ova. All of these parts are always internal. The vagina is attached to the uterus through the cervix, while the uterus is attached to the ovaries via the Fallopian tubes. At certain intervals, the ovaries release an ovum, which passes through the fallopian tube into the uterus.

If, in this transit, it meets with sperm, the sperm penetrate and merge with the egg, fertilizing it. The fertilization usually occurs in the oviducts, but can happen in the uterus itself. The zygote then implants itself in the wall of the uterus, where it begins the processes of embryogenesis and morphogenesis. When developed enough to survive outside the womb, the cervix dilates and contractions of the uterus propel the fetus through the birth canal, which is the vagina.

The ova, which are the female sex cells, are much larger than the sperm and are normally formed within the ovaries of the fetus before its birth. They are mostly fixed in location within the ovary until their transit to the uterus, and contain nutrients for the later zygote and embryo. Over a regular interval, in response to hormonal signals, a process of oogenesis matures one ovum which is released and sent down the Fallopian tube. If not fertilized, this egg is flushed out of the system through menstruation in humans and other great apes and reabsorbed in other mammals in the estrus cycle.

Gestation

Gestation, called *pregnancy* in humans, is the period of time during which the fetus develops, dividing via mitosis inside the female. During this time, the fetus receives all of its nutrition and oxygenated blood from the female, filtered through the placenta, which is attached to the fetus' abdomen via an umbilical cord. This drain of nutrients can be quite taxing on the female, who is required to ingest slightly higher levels of calories. In addition, certain vitamins and other nutrients are required in greater quantities than normal, often creating abnormal eating habits. The length of gestation, called the gestation period, varies greatly from species to species; it is 40 weeks in humans, 56–60 in giraffes and 16 days in hamsters.

Birth

Once the fetus is sufficiently developed, chemical signals start the process of birth, which begins with contractions of the uterus and the dilation of the cervix. The fetus then descends to the cervix, where it is pushed out into the vagina, and eventually out of the female. The newborn, which is called an infant in humans, should typically begin respiration on its own shortly after birth. Not long after, the placenta is passed as well. Most mammals eat this, as it is a good source of protein and other vital nutrients needed for caring for the young. The end of the umbilical cord attached to the young's abdomen eventually falls off on its own.

Monotremes

Monotremes, only five species of which exist, all from Australia and New Guinea, lay eggs. They have one opening for excretion and reproduction called the cloaca. They hold the eggs internally for several weeks, providing nutrients, and then lay them and cover them like birds. After less than two weeks the young hatches and crawls into its mother's pouch, much like marsupials, where it nurses for several weeks as it grows.

Marsupials

Marsupials reproductive systems differ markedly from those of placental mammals. Females have two vaginas, both of which open externally through one orifice but lead to different compartments within the uterus. Males generally have a two-pronged penis, which corresponds to the females' two vaginae. The penis is used only for discharging semen into females, and is separate from the urinary tract. Both sexes possess a cloaca, which is connected to a urogenital sac used to store waste before expulsion.

The female develops a kind of yolk sack in her womb which delivers nutrients to the embryo. Embryos of bandicoots, koalas and wombats additionally form placenta-like organs that connect them to the uterine wall, although the placenta-like organs are smaller than in placental mammals and it is not certain that they transfer nutrients from the mother to the embryo.

Pregnancy is very short, typically 4 to 5 weeks. The embryo is born at a very young stage of development, and is usually less than 5 cm (2.0 in) long at birth. It has been suggested that the short pregnancy is necessary to reduce the risk that the mother's immune system will attack the embryo.

The newborn marsupial uses its forelimbs (with relatively strong hands) to climb to a nipple, which is usually in a pouch on the mother's belly. The mother feeds the baby by contracting muscles over her mammary glands, as the baby is too weak to suckle. The newborn marsupial's need to use its forelimbs for climbing to the nipple has prevented the forelimbs from evolving into paddles or wings and has therefore prevented the appearance of aquatic or truly flying marsupials (although there are several marsupial gliders).

Fish

The vast majority of fish species lay eggs that are then fertilized by the male, some species lay their eggs on a substrate like a rock or on plants, while others scatter their eggs and the eggs are fertilized as they drift or sink in the water column. Some fish species use internal fertilization and then disperse the developing eggs or give birth to live offspring.

Fish that have live-bearing offspring include the Guppy and Mollies or *Poecilia.* Fishes that give birth to live young can be ovoviviparous, where the eggs are fertilized within the female and the eggs simply hatch within the female body, or in seahorses, the male carries the developing young within a pouch, and gives birth to live young. Fishes can also be viviparous, where the female supplies nourishment to the internally growing offspring. Some fish are hermaphrodites, where a single fish is both male and female and can produce eggs and sperm.

In hermaphroditic fish, some are male and female at the same time while in other fish they are serially hermaphroditic; starting as one sex and changing to the other. In at least one hermaphroditic species, self-fertilization occurs when the eggs and sperm are released together. Internal self-fertilization may occur in some other species. One fish species does not reproduce by sexual reproduction but uses sex to produce offspring; *Poecilia formosa* is a unisex species that uses a form of parthenogenesis called gynogenesis, where unfertilized eggs develop into embryos that produce female offspring. *Poecilia formosa* mate with males of other fish species that use internal fertilization, the sperm does not fertilize the eggs but stimulates the growth of the eggs which develops into embryos.

Implantation

Implantation is an event that occurs early in pregnancy in which the embryo adheres to the wall of uterus. At this stage of prenatal development, the embryo is a blastocyst. It is by this adhesion that the fetus receives the oxygen and the nutrients from the mother to be able to grow.

The Implantation Window

There are many conditions that must be satisfied in order for a successful implantation to take place. There is only a specific period of time during which implantation is possible, this is the "implantation window". A reason for this window is that if implantation does not occur at a certain time, then it signifies that something is wrong. And when there is a risk that something is wrong, there will most likely be a miscarriage rather than the continued gestation of a malformed fetus.

The implantation window is started by preparations in the endometrium of the uterus, both structurally and in the composition of its secretions.

Adaption of Uterus

To enable implantation, the uterus goes through changes in order to be able to receive the embryo.

Predecidualization

The endometrium increases in thickness, becomes more vascularized and its glands grow to be tortuous and boosted in their secretions. These changes reach their maximum about 7 days after ovulation.

Furthermore, the surface of the endometrium produces a kind of rounded cells, which cover the whole area toward the uterine cavity. This happens about 9 to 10 days after ovulation. These cells are called decidual cells, which emphasises that the whole layer of them is shed off in every menstruation if no pregnancy occurs, just as leaves of deciduous trees. The uterine glands, on the other hand, decrease in activity and degenerate already 8 to 9 days after ovulation in absence of pregnancy.

The stromal cells originate from the stromal cells that are always present in the endometrium. However, the decidual cells make up a new layer, the decidua. The rest of the endometrium, in addition, expresses differences between the luminal and the basal sides. The luminal cells form the zona compacta of the endometrium, in contrast to the basalolateral zona spongiosa, which consists of the rather spongy stromal cells.

Decidualization

Decidualization succeeds predecidualization if pregnancy occurs. This is an expansion of it, further developing the uterine glands, the zona compacta and the epithelium of decidual cells lining it. The decidual cells become filled with lipids and glycogen and take the polyhedral shape characteristic for decidual cells.

Trigger

It is likely that the blastocyst itself makes the main contribution to this additional growing and sustaining of the decidua. An indication of this is that decidualization occurs at a higher degree in conception cycles than in nonconception cycles. Furthermore, similar changes are observed when giving stimuli mimicking the natural invasion of the embryo.

Parts of Decidua

The decidua can be organized into separate sections, although they have the same composition.

Decidua Basalis

This is the part of the decidua which is located basalolateral to the embryo after implantation.

Decidua Capsularis

Decidua capsularis grows over the embryo on the luminal side, enclosing it into the endometrium. It surrounds the embryo together with decidua basalis.

Decidua Vera

All other decidua on the uterine surface belongs to decidua parietalis.

Decidua Throughout Pregnancy

After implantation the decidua remains, at least through the first trimester. However, its most prominent time is during the early stages of pregnancy, during implantation. Its function as a surrounding tissue is replaced by the definitive placenta.

However, some elements of the decidualization remain throughout pregnancy. The compacta and spongiosa layers are still observable beneath the decidua in pregnancy.

The glands of the spongiosa layer continue to secrete during the first trimester, when they degenerate. However, before that disappearance, some glands secrete unequally much. This phenomenon of hypersecretion is called the Arias-Stella phenomenon, after the pathologist Javier Arias-Stella.

Pinopodes

Pinopodes are small, finger-like protrusions from the endometrium. They appear between day 19 and day 21 of gestational age. This corresponds to a fertilization age of approximately 5 to 7 days, which corresponds well with the time of implantation. They only persist for 2 to 3 days. The development of them is enhanced by progesterone but inhibited by estrogens.

Function in Implantation

Pinopodes endocytose uterine fluid and macromolecules in it. By doing so, the volume of the uterus decreases, taking the walls closer to the embryoblast floating in it. Thus, the period of active pinocytes might also limit the implantation window.

Function During Implantation

Pinopodes continue to absorb fluid, and removes most of it during the early stages of implantation.

Adaption of Secretions

Not only the lining of the uterus transforms. In addition, the secretion from its epithelial glands changes. This change is induced by increased levels of progesterone from the corpus luteum. The target of the secretions is the embryoblast, and has several functions on it.

Nourishment

The embryoblast spends approximately 72 hours in the uterine cavity before implanting. In that time, it cannot receive nourishment directly from the blood of the mother, and must rely on secreted nutrients into the uterine cavity, e.g. iron and fat-soluble vitamins.

Growth and Implantation

In addition to nourishment, the endometrium secretes several steroid-dependent proteins, important for growth and implantation. Cholesterol and steroids are also secreted. Implantation is further facilitated by synthesis of matrix substances, adhesion molecules and surface receptors for the matrix substances.

Mechanism

Implantation occurs approximately 7 days after fertilisation, and is initiated when the blastocyst comes into contact with the uterine wall.

Zona Hatching

To be able to perform implantation, the blastocyst first needs to get rid of its zona pellucida. This process can be called "hatching".

Factors

Lytic factors in the uterine cavity, as well as factors from the blastocyst itself are essential for this process. Mechanisms in the latter are indicated by that the zona pellucida remains intact if an unfertilized egg is placed in the uterus under the same conditions. A substance probably involved is plasmin.

Plasminogen, the plasmin precursor, is found in the uterine cavity, and blastocyst factors contribute to its conversion to active plasmin. This hypothesis is supported by lytic effects in vitro by plasmin. Furthermore, plasmin inhibitors also inhibit the entire zona hatching in rat experiments.

Apposition

The very first, albeit loose, connection between the blastocyst and the endometrium is called the apposition.

Location

On the endometrium, the apposition is usually made where there is a small crypt in it, perhaps because it increases the area of contact with the rather spherical blastocyst.

On the blastocyst, on the other hand, it occurs at a location where there has been enough lysis of the zona pellucida to have created a rupture to enable direct contact between the underlying trophoblast and the decidua of the endometrium.

However, ultimately, the inner cell mass, inside the trophoblast layer, is aligned closest to the decidua. Nevertheless, the apposition on the blastocyst is not dependent on if it is on the same side of the blastocyst as the inner cell mass.

Rather, the inner cell mass rotates inside the trophoblast to align to the apposition. In short, the entire surface of the blastocyst has a potential to form the apposition to the decidua.

Adhesion

Adhesion is a much stronger attachment to the endometrium than the loose apposition.

The trophoblasts adhere by penetrating the endometrium, with protrusions of trophoblast cells.

Communication

There is massive communication between the blastocyst and the endometrium at this stage. The blastocyst signals to the endometrium to adapt further to its presence, e.g. by changes in the cytoskeleton of decidual cells. This, in turn, dislodges the decidual cells from their connection to the underlying basal lamina, which enables the blastocyst to perform the succeeding invasion.

This communication is conveyed by receptor-ligand-interactions, both integrin-matrix and proteoglycan ones.

Proteoglycan Receptors

Another ligand-receptor system involved in adhesion is proteoglycan receptors, found on the surface of the decidua of the uterus. Their counterparts, the proteoglycans, are found around the trophoblast cells of the blastocyst. This ligand-receptor system also is present just at the implantation window.

Invasion

Invasion is an even further establishment of the blastocyst in the endometrium.

Syncytiotrophoblasts

The protrusions of trophoblast cells that adhere into the endometrium continue to proliferate and penetrate into the endometrium. These penetrating cells differentiate to become a new type of cells, syncytiotrophoblast.

The prefix syn refers to that the boundaries between these cells disappears, forming a single mass of a multitude of cell nuclei. The rest of the trophoblasts, surrounding the inner cell mass, are hereafter called cytotrophoblasts.

Invasion continues with the syncytiotrophoblasts reaching the basal membrane beneath the decidual cells, penetrating it and further invading into the uterine stroma. Finally, the whole embryo is embedded in the endometrium. Eventually, the syncytiotrophoblasts come into contact with maternal blood and form chorionic villi. This is the initiation of forming the placenta.

Secretions

The blastocyst secretes factors for a multitude of purposes during invasion. It secretes several autocrine factors, targeting itself and stimulating it to further invade the endometrium. Furthermore, secretions loosen decidual cells from each other, prevent the embryo from being rejected by the mother, trigger the final decidualization and prevent menstruation.

Autocrine

Human chorionic gonadotropin is an autocrine growth factor for the blastocyst. Insulin-like growth factor type 2, on the other hand, stimulates the invasiveness of it.

Dislodging

The syncytiotrophoblasts dislodges decidual cells in their way, both by degradation of cell adhesion molecules linking the decidual cells together as well as degradation of the extracellular matrix between them.

Cell adhesion molecules are degraded by syncytiotrophoblast secretion of Tumor necrosis factor-alpha. This inhibits the expression of cadherins and beta-catenin. Cadherins is a cell adhesion molecule and beta-catenin helps to anchor it to the cell membrane. Inhibited expression of these molecules thus loosens the connection between decidual cells, permitting the syncytotrophoblasts and the whole embryo with them to invade into the endometrium. The extracellular matrix is degraded by serine endopeptidases and metalloproteinases. Examples of such metalloproteinases are collagenases, gelatinases and stromelysins. These collagenases digest Type-I collagen, Type-II collagen, Type-III collagen, Type-VII collagen and Type-X collagen. The gelatinases exist in two forms; one digesting Type-IV collagen and one digesting gelatin.

Immunosuppressive

The embryo differs from the cells of the mother, and would be rejected as a parasite by the immune system of the mother if it didn't secrete immunosuppressive agents. Such agents are Platelet-activating factor, human chorionic gonadotropin, early pregnancy factor, immunosuppressive factor, Prostaglandin E2, Interleukin 1-alpha, Interleukin 6, interferon-alpha, leukemia inhibitory factor and Colony-Stimulating Factor.

Decidualization

Factors from the blastocyst also trigger the final formation of decidual cells into their proper form. In contrast, some decidual cells in the proximity of the blastocyst degenerate, providing nutrients for it.

Prevention of Menstruation

Human chorionic gonadotropin (hCG) not only acts as an immunosuppressive, but also "notifies" the mother's body that she is pregnant, preventing menstruation by sustaining the function of the corpus luteum.

Other Factors

Other factors secreted by the blastocyst are;

- ovum factor
- Embryo-derived histamine-releasing factor
- Tissue plasminogen activator as well as its inhibitors
- Estradiol
- â1-integrins
- Fibroblast growth factor
- Transforming growth factor alpha
- inhibin.

Mammals

Usually mammals rely on internal fertilisation through copulation. After a male ejaculates, a large number of sperm cells move to the upper vagina (via contractions from the vagina) through the cervix and across the length of the uterus toward the ovum. The capacitated spermatozoon and the oocyte meet and interact in the *ampulla* of the fallopian tube. It is probable that chemotaxis is involved in directing the sperm to the egg, but the mechanism has yet to be worked out. However, demonstration of formyl peptide receptors (60.000 receptor/cell; higher binding capacity in the tail region) in the surface membrane of human sperms strongly supports, that-besides specific chemoattractant substances i.e. resact-professional chemoattractant ligands like formyl Met-Leu-Phe (fMLF) have also the ability to induce migration of sperm. The sperm binds to the zona pellucida of the egg. In contrast to sea urchins, the sperm binds to the egg before the acrosmal reaction. The zona pellucida is a thick layer of extracellular matrix that surrounds the egg and is similar to the role of the vitelline membrane in sea urchins. A glycoprotein in the zona pellucida, ZP3 was discovered to be responsible for egg/sperm adhesion *in mice.* The receptor galactosyl transferase (GalT) binds to the N-acetylglucosamine residues on the ZP3 and is important for binding to sperm and activating the acrosome reaction. ZP3 is sufficient for sperm/egg binding but not necessary. There are two additional sperm receptors: a 250kD protein that binds to an oviduct secreted protein and SED1 which binds independently to the zona. After the acrosome reaction, it is believed that the sperm remains bound to the zona pellucida through exposed ZP2 receptors. These receptors are unknown in mice but have been identified in guinea pigs. In mammals, binding of the spermatozoon to the GalT initiates the acrosome reaction. This process releases the enzyme hyaluronidase, which digests the matrix of hyaluronic acid in the vestments surrounding the oocyte. Fusion between the sperm and oocyte plasma membranes follows, allowing the entry of the sperm nucleus, centriole and flagellum, but not the mitochondria, into the oocyte. The fusion is likely mediated by the protein CD9 in mice (the binding homolog). The egg "activates" once it fuses with a single sperm cell, *i.e.*, its cell membrane changes to preventing fusion with other sperm. This process ultimately leads to the formation of a diploid cell called a zygote. The zygote begins to divide and form a blastocyst and when it reaches the uterus, it performs implantation in the endometrium. At this point the female is said to be pregnant. If the embryo emplants in any tissue other than the uterine wall, an ectopic pregnancy results, which can be fatal to the mother.

In some animals (e.g. rabbits) the act of coitus induces ovulation by stimulating release of the pituitary hormone gonadotropin. This greatly increases the probability that coitus will result in pregnancy.

Humans

The term *conception* commonly refers to fertilisation, the successful fusion of gametes to form a new organism. 'Conception' is used by some to refer to implantation and is thus a subject of semantic arguments about the beginning of pregnancy, within the abortion debate. Gastrulation, which occurs around 16 days after fertilisation, is the point in development when the implanted blastocyst develops three germ layers, the endoderm, the ectoderm and the mesoderm. It is at this point that the genetic code of the father becomes fully involved in the development of the embryo.

Until this point in development, twinning is possible. Additionally, interspecies hybrids survive only until gastrulation, and have no chance of development afterward. However this stance is not entirely accepted as some human developmental biology literature refers to the "conceptus" and such medical literature refers to the "products of conception" as the post-implantation embryo and its surrounding membranes. The term "conception" is not usually used in scientific literature because of its variable definition and connotation.

Fertilisation and Genetic Recombination

Meiosis results in a random segregation of the genes contributed from each parent. Each parent organism generally has the same genetic make-up, but differs for a fraction of their genes. Therefore, each gamete produced by a person will be genetically different from the others from that person, as well as from the gametes produced by another person.

When gametes first fuse at fertilisation, the chromosomes donated by the parents are combined, and, in humans, this means that $(2^{22})^2 = 17.6 \times 10^{12}$ chromosomally different zygotes are possible for the non-sex chromosomes, even assuming no chromosomal crossover.

If crossover occurs once, then on average $(4^{22})^2 = 309 \times 10^{24}$ genetically different zygotes are possible for every couple, not considering that crossover events can take place at most points along each chromosome. The X and Y chromosomes do not undergo crossover events, so are excluded from the calculation. Note that the mitochondrial DNA is only inherited from the maternal parent.

Parthenogenesis

Another method of fertilisation occurs among animals that normally reproduce sexually, through parthenogenesis: when the gamete of a female is not fertilised by a male, yet produces viable and unique offspring that are not clones. Only DNA from the mother is inherited, but it is not identical to her.

Normal eggs of the mother become fertilised, without sperm, and development proceeds normally. This occurs naturally in several species and may be induced in others through a chemical or electrical stimulus. In 2004, Japanese researchers led by Tomohiro Kono succeeded after 457 attempts to merge the ova of two mice, the result of which developed normally into a mouse. This was achieved by blocking certain proteins that would normally prevent the possibility.

Oviduct

In non-mammalian vertebrates, the passage from the ovaries to the outside of the body is known as the oviduct. The eggs travel along the oviduct. These eggs will either be fertilized by sperm to become a zygote, or will degenerate in the body. Normally, these are paired structures, but in birds, crocodilians, and some cartilaginous fishes, one or the other side fails to develop (together with the corresponding ovary), and only one functional oviduct is found.

Except in teleosts, the oviduct does not directly contact the ovary. Instead, the most anterior portion ends in a funnel-shaped structure called the infundibulum, which collects eggs as they are released by the ovary into the body cavity.

The only female vertebrates to lack oviducts are the jawless fishes. In these species, the single fused ovary releases eggs directly into the body cavity. The fish eventually extrudes the eggs through small genital pores towards the rear of the body.

Fish and Amphibians

In amphibians and lungfishes, the oviduct is a simple ciliated tube, lined with mucus-secreting glands that produce the jelly that surrounds the ovum. In all other vertebrates, there is normally some degree of specialisation of the tube, depending on the type of eggs produced.

In cartilaginous fishes, the mid portion of the tube develops as a shell gland. The first portion of this gland secretes the egg white, while the lower portion secretes a hard, horny, capsule to protect the developing egg. Below the shell gland is the ovisac, a distended region in which eggs are stored prior to laying. In ovoviviparous species, the egg remains within the ovisac until it hatches. Some cartilaginous fishes, however, are truly viviparous, giving birth to live young, and producing no egg shell. In these forms the ovisac nurtures the developing embryo, often with the aid of vascular outgrowths similar to, but much simpler than, the mammalian placenta.

The most primitive ray-finned fishes retain the simple structure also found in lungfishes, but in teleosts, folds of peritoneum enclose the ovary and upper part of the tube, fusing them into a single structure. The ovary itself is hollow, with eggs being shed into the central cavity, and thence passing directly into the oviduct. The enclosed nature of the female reproductive system in these fishes makes it impossible for eggs to escape into the general body cavity; a necessary development given that thousands or even millions of eggs may be released in a single spawning.

Amniotes

In amniotes-reptiles, birds, and mammals-the egg is enclosed with an outer layer, or amnion, which has led to further development of the oviduct. In reptiles, birds, and monotremes, the main part of the oviduct is a muscular tube, capable of considerable distension to transport the large eggs that are produced. This part of the oviduct is lined with glands that secrete the components of the egg white. The lower portion of the oviduct, or uterus, has a thicker layer of smooth muscle and contains the glands that secrete the egg shell. In marsupials and placental mammals, the uterus becomes lined by an endometrium, and is more developed than in egg-laying amniotes. In many placental mammals, the uteri of each side become partially or wholly fused into a single organ, although in marsupials they remain completely separate. In mammals, the portion of the oviduct above the uterus is referred to as the fallopian tube.

In birds

The oviduct in birds is divided in several parts:

- *Infundibulum* (formation of *Chalazae*, place of fertilisation)
- *Magnum* (formation of egg white)
- *Isthmus* (formation of the shell membrane)
- *Shell Gland* (formation of egg shell)
- *Vagina* (formation of cuticula).

Oviparity

Oviparous animals are animals that lay eggs, with little or no other embryonic development within the mother. This is the reproductive method of most fish, amphibians, reptiles, all birds, the monotremes, and most insects and arachnids. Land-dwelling animals that lay eggs, often protected by a shell, such as reptiles and insects, do so after having completed the process of internal fertilization. Water-dwelling animals, such as fish and amphibians, lay their eggs before fertilization, and the male lays its sperm on top of the newly laid eggs in a process called external fertilization. Almost all non-oviparous fish, amphibians and reptiles are ovoviviparous, i.e. the eggs are hatched inside the mother's body (or, in case of the sea horse inside the father's). The true opposite of oviparity is placental viviparity, employed by almost all mammals (the exceptions being marsupials and monotremes).

There are only five known species of oviparous mammals: four species of Echidna and the Platypus.

Endometrium

Function

The endometrium functions as a lining for the uterus, preventing adhesions between the opposed walls of the myometrium, thereby maintaining the patency of the uterine cavity. During the menstrual cycle or estrous cycle, the endometrium grows to a thick, blood vessel-rich, glandular tissue layer. This represents an optimal environment for the implantation of a blastocyst upon its arrival in the uterus.

During pregnancy, the glands and blood vessels in the endometrium further increase in size and number. Vascular spaces fuse and become interconnected, forming the placenta, which supplies oxygen and nutrition to the embryo and fetus.

Cycle

The endometrial lining undergoes cyclic regeneration. Humans and the great apes display the menstrual cycle, whereas most other mammals are subject to an estrous cycle. In both cases, the endometrium initially proliferates under the influence of estrogen. However, once ovulation occurs, in addition to estrogen, the ovary will also start to produce progesterone. This changes the proliferative pattern of the endometrium to a secretory lining. Eventually, the secretory lining provides a hospitable environment for one or more blastocysts.

If no blastocyst is detected, the progesterone level drops and the endometrial lining is either reabsorbed (estrous cycle) or shed (menstrual cycle). In the latter case, the process of shedding involves the breaking down of the lining, the tearing of small connective blood vessels, and the loss of the tissue and blood that had constituted it through the vagina. The entire process occurs over a period of several days. Menstruation may be accompanied by a series of uterine contractions; These help expel the menstrual endometrium.

In case of implantation, however, the endometrial lining is neither absorbed nor shed. Instead, it remains as *decidua*. The decidua becomes part of the placenta; it provides support and protection for the gestation.

If there is inadequate stimulation of the lining, due to lack of hormones, the endometrium remains thin and inactive. In humans, this will result in amenorrhea. After menopause, the lining is often described as being atrophic. In contrast, endometrium that is chronically exposed to estrogens, but not to progesterone, may become hyperplastic. In humans, the cycle of building and shedding the endometrial lining lasts an average of 28 days. The endometrium develops at different rates in different mammals. Its formation is sometimes affected by seasons, climate, stress, and other factors. The endometrium itself produces certain hormones at different points along the cycle. This affects other portions of the reproductive system.

Histology

The endometrium consists of a single layer of columnar epithelium, resting on a layer of connective tissue, which varies in thickness according to hormonal influences-the *stroma.* Simple tubular uterine glands reach from the endometrial surface through to the base of the stroma, which also carries a rich blood supply of spiral arteries. In a woman of reproductive age, two layers of endometrium can be distinguished. These two layers occur only in endometrium lining the cavity of the uterus, not in the lining of the Fallopian tubes:

- The *functional layer* is adjacent to the uterine cavity. This layer is built up after the end of menstruation during the first part of the previous menstrual cycle. Proliferation is induced by estrogen (follicular phase of menstrual cycle), and later changes in this layer are engendered by progestrone from the corpus luteum (luteal phase). It is adapted to provide an optimum environment for the implantation and growth of the embryo. This layer is completely shed during menstruation.
- The *basal layer*, adjacent to the myometrium and below the functional layer, is not shed at any time during the menstrual cycle, and from it the functional layer develops.

In the absence of progesterone, the arteries supplying blood to the functional layer constrict, so that cells in that layer become ischaemic and die, leading to menstruation.

It is possible to identify the phase of the menstrual cycle by observing histological differences at each phase:

Phase	***Days***	***Thickness***	***Epithelium***
menstrual phase	1-4	thin	absent
proliferative phase	4-14	intermediate	columnar
secretory phase	15-28	thick	columnar. Also visible are helicine branches of uterine artery

Pathological Conditions

Adenomyosis is the growth of the endometrium into the muscle layer of the uterus (the myometrium).

Endometriosis is the growth of endometrial tissue outside the uterus.

Endometrial cancer is the most common cancer of the human female genital tract.

Asherman's syndrome, also known as intrauterine adhesions occurs when the basal layer of the endometrium is damaged by instrumention (eg. D&C) or infection (eg. endometrial tuberculosis) resulting in endometrial sclerosis and adhesion formation partially or completely obliterating the uterine cavity.

Mitosis

Mitosis is the process by which a eukaryotic cell separates the chromosomes in its cell nucleus into two identical sets in two nuclei. It is generally followed immediately by cytokinesis, which divides the nuclei, cytoplasm, organelles and cell membrane into two cells containing roughly equal shares of these cellular components. Mitosis and cytokinesis together define the mitotic (M) phase of the cell cycle- the division of the mother cell into two daughter cells, genetically identical to each other and to their parent cell. This accounts for approximately 10% of the cell cycle.

Mitosis occurs exclusively in eukaryotic cells, but occurs in different ways in different species. For example, animals undergo an "open" mitosis, where the nuclear envelope breaks down before the chromosomes separate, while fungi such as *Aspergillus nidulans* and *Saccharomyces cerevisiae* (yeast) undergo a "closed" mitosis, where chromosomes divide within an intact cell nucleus. Prokaryotic cells, which lack a nucleus, divide by a process called binary fission.

The process of mitosis is complex and highly regulated. The sequence of events is divided into phases, corresponding to the completion of one set of activities and the start of the next. These stages are prophase, prometaphase, metaphase, anaphase and telophase. During the process of mitosis the pairs of chromosomes condense and attach to fibers that pull the sister chromatids to opposite sides of the cell. The cell then divides in cytokinesis, to produce two identical daughter cells.

Because cytokinesis usually occurs in conjunction with mitosis, "mitosis" is often used interchangeably with "mitotic phase". However, there are many cells where mitosis and cytokinesis occur separately, forming single cells with multiple nuclei.

This occurs most notably among the fungi and slime moulds, but is found in various different groups. Even in animals, cytokinesis and mitosis may occur independently, for instance during certain stages of fruit fly embryonic development. Errors in mitosis can either kill a cell through apoptosis or cause mutations that may lead to cancer.

Overview

The primary result of mitosis is the division of the parent cell's genome into two daughter cells. The genome is composed of a number of chromosomes-complexes of tightly-coiled DNA that contain genetic information vital for proper cell function. Because each resultant daughter cell should be genetically identical to the parent cell, the parent cell must make a copy of each chromosome before mitosis. This occurs during S phase, in interphase, the period that precedes the mitotic phase in the cell cycle where preparation for mitosis occurs.

Each new chromosome now contains two identical copies of itself, called *sister chromatids*, attached together in a specialized region of the chromosome known as the *centromere*. Each sister chromatid is not considered a chromosome in itself, and a chromosome does not always contain two sister chromatids. In most eukaryotes, the nuclear envelope that separates the DNA from the cytoplasm disassembles. The chromosomes align themselves in a line spanning the cell. Microtubules, essentially miniature strings, splay out from opposite ends of the cell and shorten, pulling apart the sister chromatids of each chromosome. As a matter of convention, each sister chromatid is now considered a chromosome, so they are renamed to *sister chromosomes*. As the cell elongates, corresponding sister chromosomes are pulled toward opposite ends. A new nuclear envelope forms around the separated sister chromosomes.

As mitosis completes cytokinesis is well underway. In animal cells, the cell pinches inward where the imaginary line used to be (the pinching of the cell membrane to form the two daughter cells is called cleavage furrow), separating the two developing nuclei. In plant cells, the daughter cells will construct a new dividing cell wall between each other. Eventually, the mother cell will be split in half, giving rise to two daughter cells, each with an equivalent and complete copy of the original genome.

Prokaryotic cells undergo a process similar to mitosis called binary fission. However, prokaryotes cannot be properly said to undergo mitosis because they lack a nucleus and only have a single chromosome with no centromere.

Phases of Cell Cycle and Mitosis

Interphase

The mitotic phase is a relatively short period of the cell cycle. It alternates with the much longer *interphase*, where the cell prepares itself for cell division. Interphase is therefore not part of mitosis. Interphase is divided into three phases, G_1 (first gap), S (synthesis), and G_2 (second gap). During all three phases, the cell grows by producing proteins and cytoplasmic organelles. However, chromosomes are replicated only during the S phase. Thus, a cell grows (G_1), continues to grow as it duplicates its chromosomes (S), grows more and prepares for mitosis (G_2), and finally divides (M) before restarting the cycle.

Preprophase

In plant cells only, prophase is preceded by a pre-prophase stage. In highly vacuolated plant cells, the nucleus has to migrate into the center of the cell before mitosis can begin. This is achieved through the formation of a phragmosome, a transverse sheet of cytoplasm that bisects the cell along the future plane of cell division. In addition to phragmosome formation, preprophase is characterized by the formation of a ring of microtubules and actin filaments (called preprophase band) underneath the plasma membrane around the equatorial plane of the future mitotic spindle. This band marks the position where the cell will eventually divide. The cells of higher plants (such as the flowering plants) lack centrioles: with microtubules forming a spindle on the surface of the nucleus and then being organized into a spindle by the chromosomes themselves, after the nuclear membrane breaks down. The preprophase band disappears during nuclear envelope disassembly and spindle formation in prometaphase. Normally, the genetic material in the nucleus is in a loosely bundled coil called chromatin. At the onset of prophase, chromatin condenses together into a highly ordered structure called a chromosome. Since the genetic material has already been duplicated earlier in S phase, the replicated chromosomes have two sister chromatids, bound together at the centromere by the cohesion complex. Chromosomes are visible at high magnification through a light microscope. Close to the nucleus are structures called centrosomes, which are made of a pair of centrioles. The centrosome is the coordinating center for the cell's microtubules. A cell inherits a single centrosome at cell division, which replicates before a new mitosis begins, giving a pair of centrosomes. The two centrosomes nucleate microtubules (which may be thought of as cellular ropes or poles) to form the spindle by polymerizing soluble tubulin. Molecular motor proteins then push the centrosomes along these microtubules to opposite sides of the cell. Although centrioles help organize microtubule assembly, they are not essential for the formation of the spindle, since they are absent from plants, and centrosomes are not always used in meiosis.

Prometaphase

The nuclear envelope disassembles and microtubules invade the nuclear space. This is called open mitosis, and it occurs in most multicellular organisms. Fungi and some protists, such as algae or trichomonads, undergo a variation called closed mitosis where the spindle forms inside the nucleus or its microtubules are able to penetrate an intact nuclear envelope. Each chromosome forms two kinetochores at the centromere, one attached at each chromatid. A kinetochore is a complex protein structure that is analogous to a ring for the microtubule hook; it is the point where microtubules attach themselves to the chromosome. Although the kinetochore structure and function are not fully understood, it is known that it contains some form of molecular motor. When a microtubule connects with the kinetochore, the motor activates, using energy from ATP to "crawl" up the tube toward the originating centrosome. This motor activity, coupled with polymerisation and depolymerisation of microtubules, provides the pulling force necessary to later separate the chromosome's two chromatids. When the spindle grows to sufficient length, *kinetochore microtubules* begin searching for kinetochores to attach to. A number of *nonkinetochore microtubules* find and interact with corresponding nonkinetochore microtubules from the opposite centrosome to form the mitotic spindle. Prometaphase is sometimes considered part of prophase.

In the fishing pole analogy, the kinetochore would be the "hook" that catches a sister chromatid or "fish". The centrosome acts as the "reel" that draws in the spindle fibers or "fishing line".

Metaphase

As microtubules find and attach to kinetochores in prometaphase, the centromeres of the chromosomes convene along the *metaphase plate* or *equatorial plane*, an imaginary line that is equidistant from the two centrosome poles. This even alignment is due to the counterbalance of the pulling powers generated by the opposing kinetochores, analogous to a tug-of-war between people of equal strength. In certain types of cells, chromosomes do not line up at the metaphase plate and instead move back and forth between the poles randomly, only roughly lining up along the midline. Metaphase comes from the Greek *Iowa* meaning "after." Because proper chromosome separation requires that every kinetochore be attached to a bundle of microtubules (spindle fibres), it is thought that unattached kinetochores generate a signal to prevent premature progression to anaphase without all chromosomes being aligned. The signal creates the *mitotic spindle checkpoint.*

Anaphase

When every kinetochore is attached to a cluster of microtubules and the chromosomes have lined up along the metaphase plate, the cell proceeds to anaphase. Two events then occur; First, the proteins that bind sister chromatids together are cleaved, allowing them to separate. These sister chromatids, which have now become distinct sister chromosomes, are pulled apart by shortening kinetochore microtubules and move toward the respective centrosomes to which they are attached. Next, the nonkinetochore microtubules elongate, pulling the centrosomes (and the set of chromosomes to which they are attached) apart to opposite ends of the cell. The force that causes the centrosomes to move towards the ends of the cell is still unknown, although there is a theory that suggests that the rapid assembly and breakdown of microtubules may cause this movement. These two stages are sometimes called early and late anaphase. Early anaphase is usually defined as the separation of the sister chromatids, while late anaphase is the elongation of the microtubules and the chromosomes being pulled farther apart. At the end of anaphase, the cell has succeeded in separating identical copies of the genetic material into two distinct populations.

Telophase

Telophase is a reversal of prophase and prometaphase events. It "cleans up" the after effects of mitosis. At telophase, the nonkinetochore microtubules continue to lengthen, elongating the cell even more. Corresponding sister chromosomes attach at opposite ends of the cell. A new nuclear envelope, using fragments of the parent cell's nuclear membrane, forms around each set of separated sister chromosomes. Both sets of chromosomes, now surrounded by new nuclei, unfold back into chromatin. Mitosis is complete, but cell division is not yet complete.

Cytokinesis

Cytokinesis is often mistakenly thought to be the final part of telophase; however, cytokinesis is a separate process that begins at the same time as telophase. Cytokinesis is technically not even a phase of mitosis, but rather a separate process, necessary for completing cell division. In animal cells, a cleavage furrow (pinch) containing a contractile ring develops where the metaphase plate used to be, pinching off the separated nuclei. In both animal and plant cells, cell division is also driven by vesicles derived from the Golgi apparatus, which move along microtubules to the middle of the cell. In plants this structure coalesces into a cell plate at the center of the phragmoplast and develops into a cell wall, separating the two nuclei. The phragmoplast is a microtubule structure typical for higher plants, whereas some green algae use a phycoplast microtubule array during cytokinesis. Each daughter cell has a complete copy of the genome of its parent cell. The end of cytokinesis marks the end of the M-phase.

Significance

Mitosis is important for the maintenance of the chromosomal set; each cell formed receives chromosomes that are alike in composition and equal in number to the chromosomes of the parent cell. Transcription is generally believed to cease during mitosis, but epigenetic mechanisms such as bookmarking function during this stage of the cell cycle to ensure that the "memory" of which genes were active prior to entry into mitosis are transmitted to the daughter cells.

Consequences of Errors

Although errors in mitosis are rare, the process may go wrong, especially during early cellular divisions in the zygote. Mitotic errors can be especially dangerous to the organism because future offspring from this parent cell will carry the same disorder. In *non-disjunction*, a chromosome may fail to separate during anaphase. One daughter cell will receive both sister chromosomes and the other will receive none. This results in the former cell having three chromosomes containing the same genes (two sisters and a homologue), a condition known as *trisomy*, and the latter cell having only one chromosome (the homologous chromosome), a condition known as *monosomy*. These cells are considered aneuploid, a condition often associated with cancer. Mitosis is a traumatic process. The cell goes through dramatic changes in ultrastructure, its organelles disintegrate and reform in a matter of hours, and chromosomes are jostled constantly by probing microtubules. Occasionally, chromosomes may become damaged. An arm of the chromosome may be broken and the fragment lost, causing deletion. The fragment may incorrectly reattach to another, non-homologous chromosome, causing translocation. It may reattach to the original chromosome, but in reverse orientation, causing inversion. Or, it may be treated erroneously as a separate chromosome, causing chromosomal duplication. The effect of these genetic abnormalities depends on the specific nature of the error. It may range from no noticeable effect to cancer induction or organism death.

Endomitosis

Endomitosis is a variant of mitosis without nuclear or cellular division, resulting in cells with many copies of the same chromosome occupying a single nucleus. This process may also be referred to as endoreduplication and the cells as endoploid. An example of a cell that goes through endomitosis is the megakaryocyte.

Human Chorionic Gonadotropin

Structure

Human chorionic gonadotropin is a glycoprotein composed of 244 amino acids with a molecular mass of 36.7 kDa. Its total dimensions are 75×35×30 angstroms (7.5×3.5×3 nanometers).

It is heterodimeric, with an á (alpha) subunit identical to that of luteinizing hormone (LH), follicle-stimulating hormone (FSH), thyroid-stimulating hormone (TSH), and â (beta) subunit that is unique to hCG.

- The á (alpha) subunit is 92 amino acids long and has dimensions 60×25×15 angstroms (6×2.5×1.5 nm).
- The â-subunit of hCG gonadotropin contains 145 amino acids and has dimensions 6.5×2.5×2 nm, encoded by six highly-homologous genes that are arranged in tandem and inverted pairs on chromosome 19q13.3-*CGB* (*1, 2, 3, 5, 7, 8*).

The two subunits create a small hydrophobic core surrounded by a high surface area-to-volume ratio: 2.8 times that of a sphere. The vast majority of the outer amino acids are hydrophilic.

Function

Human chorionic gonadotropin interacts with the LHCG receptor and promotes the maintenance of the corpus luteum during the beginning of pregnancy, causing it to secrete the hormone progesterone. Progesterone enriches the uterus with a thick lining of blood vessels and capillaries so that it can sustain the growing fetus. Due to its highly-negative charge, hCG may repel the immune cells of the mother, protecting the fetus during the first trimester. It has also been hypothesized that hCG may be a placental link for the development of local maternal immunotolerance. For example, hCG-treated endometrial cells induce an increase in T cell apoptosis (dissolution of T-cells). These results suggest that hCG may be a link in the development of peritrophoblastic immune tolerance, and may facilitate the trophoblast invasion, which is known to expedite fetal development in the endometrium. It has also been suggested that hCG levels are linked to the severity of morning sickness in pregnant women.

Because of its similarity to LH, hCG can also be used clinically to induce ovulation in the ovaries as well as testosterone production in the testes. As the most abundant biological source is women who are presently pregnant, some organizations collect urine from pregnant women to extract hCG for use in fertility treatment.

Human chorionic gonadotropin also plays a role in cellular differentiation/proliferation and may activate apoptosis.

Production

Like other gonadotropins, hCG can be extracted from urine or by genetic modification. Pregnyl, Follutein, Profasi, and Novarel use the former method, derived from the urine of pregnant women. Ovidrel, on the other hand, is a product of recombinant DNA.

Testing

Levels of hCG may be measured in the blood or urine. Most commonly, this is done as a pregnancy test, intended to indicate the presence or absence of an implanted embryo. Testing for hCG may also be done when diagnosing or monitoring germ cell tumours and gestational trophoblastic disease.

Most tests employ a monoclonal antibody, which is specific to the *â-subunit* of hCG (â-hCG). This procedure is employed to ensure that tests do not make false positives by confusing hCG with LH and FSH. (The latter two are always present at varying levels in the body, whereas the presence of hCG almost always indicates pregnancy.)

- The urine test may be a *chromatographic immunoassay* or any of several other test formats, home-, physician's office-, or laboratory-based. Published detection thresholds range from 20 to 100 mIU/ml, depending on the brand of test. Early in pregnancy, more accurate results may be obtained by using the first urine of the morning when hCG levels are highest. When the urine is dilute (specific gravity less than 1.015), the hCG concentration may not be representative of the blood concentration, and the test may be falsely negative.
- The serum test, using 2-4 mL of venous blood, is typically a chemiluminescent or fluorimetric immunoassay that can detect âhCG levels as low as 5 mIU/ml and allows quantification of the âhCG concentration. The ability to quantitate the âhCG level is useful in the monitoring germ cell and trophoblastic

tumours, followup care after miscarriage, and in diagnosis of and follow-up care after treatment of ectopic pregnancy. The lack of a visible fetus on vaginal ultrasound after the âhCG levels have reached 1500 IU/ml is strongly indicative of an ectopic pregnancy.

Gestational trophoblastic disease like Hydatidiform moles ("molar pregnancy") or Choriocarcinoma may produce high levels of âhCG (due to the presence of syncytialtrophoblasts-part of the villi that make up the placenta) despite the absence of an embryo. This, as well as several other conditions, can lead to elevated hCG readings in the absence of pregnancy.

hCG levels are also a component of the triple test, a screening test for certain fetal chromosomal abnormalities/birth defects.

Uses

Tumor marker: The â subunit of human chorionic gonadotropin is secreted also by some cancers including seminoma, choriocarcinoma, germ cell tumours, hydatidiform mole formation, teratoma with elements of choriocarcinoma (this is rare), and islet cell tumor. For this reason a positive result in males can be a test for testicular cancer. The normal range for men is between 0-5 IU/ml.

Fertility: Human chorionic gonadotropin is extensively used as a parenteral fertility medication in lieu of luteinizing hormone. In the presence of one or more mature ovarian follicles, ovulation can be triggered by the administration of hCG. As ovulation will happen about 36–48 hours after the injection of hCG, procedures can be scheduled to take advantage of this time sequence. Thus, patients that undergo IVF, in general, receive hCG to trigger the ovulation process, but have their eggs retrieved at about 36 hours after injection, a few hours before the eggs actually would be released from the ovary.

As hCG supports the corpus luteum, administration of hCG is used in certain circumstances to enhance the production of progesterone. In the male, hCG injections are used to stimulate the leydig cells to synthesize testosterone. The intratesticular testosterone is necessary for spermatogenesis from the sertoli cells. Typical uses for hCG in men include hypogonadism and fertility treatment.

During first few months of pregnancy, the transmission of HIV-1 from woman to fetus is extremely rare. It has been suggested that this is due to the high concentration of hCG, and that the beta-subunit of this protein is active against HIV-1.

Weight loss: A controversial usage of hCG is as an adjunct to the British endocrinologist A.T.W. Simeons' ultra-low-calorie weight-loss diet. Simeons, while studying pregnant women in India on a calorie-deficient diet, and "fat boys" with pituitary problems treated with low-dose hCG, discovered that both lost fat rather than lean (muscle) tissue. He reasoned that hCG must be programming the hypothalamus to do this in the former cases in order to protect the developing fetus by promoting mobilization and consumption of *abnormal, excessive* adipose deposits. Simeons, practicing at Salvator Mundi International Hospital in Rome, Italy, recommended low-dose daily hCG injections (125 mg) in combination with a customized ultra-low-calorie (500 cal/day, high-protein, low-carbohydrate/fat) diet loss of adipose tissue without loss of lean tissue. After Simeons' mysterious death, the diet started to spread to specialized centers and via popularization by such as the controversial author Kevin Trudeau.

The controversy proceeds from warnings by the Journal of the American Medical Association and the American Journal of Clinical Nutrition that hCG is neither safe, nor effective as a weight-loss aid.

Anabolic Steroid Adjunct

In the world of performance enhancing drugs, hCG is increasingly used in combination with various anabolic androgenic steroid (AAS) cycles. As a result, hCG is included in some sports' illegal drug lists.

When AAS are put into a male body, the body's natural negative-feedback loops cause the body to shut down its own production of testosterone via shutdown of the hypothalamic-pituitary-gonadal axis (HPGA). This causes testicular atrophy, among other things. hCG is commonly used during and after steroid cycles to maintain and restore testicular size as well as normal testosterone production. High levels of AASs that mimic the body's natural testosterone trigger, the hypothalamus, to shut down its production of gonadotropin-releasing hormone (GnRH) from the hypothalamus. Without GnRH, the pituitary gland stops releasing luteinizing hormone (LH). LH normally travels from the pituitary via the blood stream to the testes, where it triggers the production and release of testosterone. Without LH, the testes shut down their production of testosterone. In males, hCG helps restore and maintain testosterone production in the testes by mimicking LH and triggering the production and release of testosterone.

If hCG is used for too long and in too high a dose, the resulting rise in natural testosterone will eventually inhibit its own production via negative feedback on the hypothalamus and pituitary.

Gravitational Biology of Mammalian Reproduction

Mammalian reproduction evolved within Earth's 1-g gravitational field. Therefore, deviations from Earth's normal gravity, i.e., hypogravity (forces < 1 g) or hypergravity (forces >1 g), may compromise reproduction. Some interesting findings are emerging from spaceflight studies. For example, a transient but dramatic reduction in testosterone (T) has been reported during spaceflight in male rats and humans. This observation suggests that fertility may be reduced during spaceflight, because adequate levels of T are required by adult males to maintain reproductive function. Pregnancy during space flight has been contraindicated, leading female astronauts to suppress their fertility cycles. As a result, no systematic studies have been conducted to investigate the effect of deviations from 1 g on the reproductive physiology of nonpregnant mammals. Several studies have reported no detrimental effects on pregnancy, reproductive hormones, fetal development, parturition, or lactation in female rats following return to Earth.

In the era of the International Space Station with the eventual goal of space colonization, the emphasis of space biology research has begun to shift from investigations of acute responses to studies of chronic effects of altered gravity. If space is to be successfully colonized, then a more comprehensive understanding of the effects of altered gravity on mammalian reproduction is needed. In addition, an understanding of the effects of altered gravity on reproductive physiology is imperative for the development of countermeasures. Here, we review the existing data on mammalian reproduction derived from space flight studies and from ground-based models of simulated hypo- and hypergravity. Such studies will provide the foundation to permit multiple generations of humans and other mammals to exist in space.

Ground-based Models in Gravitational Biology

There are many challenges to overcome when conducting space flight experiments with mammals. For example, it is difficult to distinguish whether reproductive changes are due to gravity or other conditions aboard the spacecraft, such as increased radiation, noise, isolation, disrupted circadian rhythms, and stress. Responses to decreased gravity on orbit versus increased gravity during landing are difficult to separate because subjects are not immediately recovered after flight. Thus, observed physiological changes may reflect recovery or landing (acute hypergravity) responses rather than in-flight (hypogravity) effects. Reproductive studies are also hampered by the short duration, high costs, and limited opportunities of space flight. To address some of these problems, scientists have designed simulation models of space flight.

One ground-based model for simulating hypogravity exposure in humans is bed rest with a 6° head-down tilt to initiate shifts in body fluids comparable to those experienced during orbital space flight. Bed rest is a convenient method that is reasonably effective at reproducing the major symptoms of hypogravity experienced during space flight. In rodent studies, the Morey-Holton hindlimb suspension (HLS) model has been used to simulate the major physiological effects of hypogravity. The HLS model involves suspending rats by the tail base to produce a 30° head-down position that complements the human 6° head-down tilt utilized in bed-rest studies.

A ground-based model for studying hypergravity is centrifugation. With the use of proper controls for Coriolis effects, centrifugation allows for an infinite number of graded increases in gravitational load and is therefore applicable to a broad range of gravity-related questions. Acute centrifugation is used to mimic the hypergravity associated with launch and landing, whereas chronic centrifugation is used to study the long-term effects of increased gravity on biological systems.

Ground-based models have been used successfully in studies of the musculoskeletal system, to validate space flight experiments, and to define adaptation to the space environment. Advantages of using ground-based models are the ability to control environmental factors more effectively than can often be achieved in space flight studies. Animals adapt well to ground-based models, thereby minimizing stress-associated physiological effects. Therefore, ground-based models are potentially valuable tools for investigating reproduction during exposure to hyper- and hypogravity.

Effects of Space flight on Male Reproduction

The testes are important organs of the male reproductive system with the dual function of spermatogenesis and steroidogenesis. Testosterone (T), the principal androgen, maintains reproductive organ function and stimulates sperm production in adult males. In most space flight studies, reduced testes weight and T levels were reported. Reduction in T following space flight is not always accompanied by reduced spermatogenesis. Reduction in T levels may not be great enough or the duration may not be long enough to affect spermatogenesis. Systematic examination of androgen-dependent accessory sex glands may provide valuable insight. However, to our knowledge no researchers have systematically examined the effects of space flight on the accessory sex glands.

In one published study, no effects on T were reported. Inconsistent results may occur because T levels in the blood fluctuate across the circadian cycle. Ortiz et al. avoiding the bias inherent in single measurement studies by collecting 24-h urinary T data, reported increased postflight urinary T in rats followed by a return to normal T levels. This finding contrasts with those of other reports of reduced T in response to space flight. The use of urine samples may provide a more comprehensive analysis of effects on the hypothalamic-pituitary-gonadal (HPG) axis. Clearly, more studies are required before definitive statements can be made regarding effects of space flight on male reproductive hormones.

Reproductive changes do not always occur at the testes but may occur at the level of the hypothalamus or pituitary. The hypothalamus produces GnRH, which acts on the pituitary gland to stimulate secretion of FSH and LH. Decreased T in various biological fluids of male astronauts occurred in parallel with increased plasma LH, which returned to normal after flight. The in-flight increase in LH was suggested to be a compensatory mechanism in response to decreased T levels. However, Ortiz et al. reported increased urinary LH in rats in response to increased urinary T after flight. In both animals and humans, the pituitary responded to changes in plasma T, indicating that the HPG axis was not impaired by space flight. Reproductive changes in males following space flight suggest that hypogravity or other factors associated with space flight may compromise male reproduction.

Effects of Simulated Hypogravity on Male Reproduction

Ground-based models used to simulate hypogravity have facilitated the interpretation of space flight studies. In ground-based simulation models, stress effects can be controlled, which has important implications for research on reproduction because stress responses are known to adversely affect reproductive function. For example, in animals and humans increased cortisol excretion associated with stress can result in reduced serum T levels. Thus, it has been difficult to ascertain whether the reproductive changes are due to stress or to hypogravity associated with space flight. Chronic stress in space flight rats was indicated by higher adrenal gland weights compared with controls. In studies of astronauts aboard the MIR Space Station, increased cortisol was reported during space flight. In ground-based studies utilizing bed rest to simulate hypogravity, Blanc et al. found elevated cortisol excretion during the initial 4 wk of bed rest. Vernikos et al. compared urinary cortisol excretion in men and women during 1 wk of bed rest and reported elevated cortisol in the men only. Subsequent studies showed no change in cortisol levels in men. In the latter study, subjects had a 15-day pre-bed-rest period during which they were permitted to adapt to the protocol and environment. Similarly, space flight studies of humans showed unchanged cortisol levels after an initial adjustment period. In humans, 6° head-down tilt bed rest for 60–120 days altered sperm morphology and reduced the number of active spermatozoa. This finding has important implications because in humans even slight reductions in sperm numbers compromise reproductive capacity. Bed rest had no effect on plasma T or LH but did produce a reduction in plasma FSH and prolactin levels. Both FSH and prolactin are involved in the regulation of spermatogenesis, suggesting that reductions in sperm number may have been due to simulated hypogravity-induced changes in endocrine function.

Rats exposed to simulated hypogravity using the Morey-Holton HLS model exhibited increased cortisol during initial HLS. No effects have been reported on stress-inducible transcripts or on adrenal gland weights, suggesting the absence of stress in HLS animals following initial adaptation. Congruent with the space flight studies, HLS rats had reduced testes weights. Longer-term (>14 days) HLS of male rats resulted in >50% reductions in testes weight compared with nonsuspended controls. Together, these results suggest that effects on testicular weight are due to simulated hypogravity rather than stress. Atrophy of testes in HLS rats was accompanied by significant reductions in plasma T compared with nonsuspended controls and an 85% loss of spermatogenic cells, indicating that both spermatogenesis and steroidogenesis were impaired in HLS rats. The results of the HLS studies of T levels have generally been difficult to interpret because of the lack of standardization across studies of animals and methods for blood collection.

Testosterone levels vary dramatically during testicular development in maturing rats and within normal diurnal cycles. In addition, studies of male reproduction using HLS require certain modifications because of the unique anatomical characteristics of rats. Unlike humans, the inguinal canal between the scrotum and abdominal cavity does not close in rats. Therefore, the 30° head-down angle used in the HLS model to unload the hindquarters can translocate the testes and epididymis into the abdomen. Exposing the testes to body cavity temperature results in infertility due to destruction of the germinal epithelium and death of spermatogonia. Histological examination of the testes of HLS rats revealed cellular damage, whereas preventing the testes from translocating into the abdomen by inguinal ligation resulted in no aberrant histological features in the testes and normal spermatogenesis. Despite inguinal ligation, T levels remained reduced in HLS rats, and despite reduced T levels, there were no significant effects on the androgen-dependent accessory sex glands of HLS rats. However, the study duration (7 days) may not have provided sufficient time for changes in the accessory sex glands to be observed.

HLS is an extremely useful model of simulated hypogravity in studies of male reproductive physiology in rats provided that care is taken to prevent translocation of the testes into the abdomen and maturational factors are taken into account. Studies using the HLS model are providing evidence that the hypogravity component of space flight exerts pronounced effects on male reproductive processes.

Effects of Hypergravity on Male Reproduction

Centrifugation studies have enabled researchers to distinguish effects of acute versus chronic hypergravity exposure. Short-term centrifugation even at high forces of 4.1 *g* for 15 min had no detectable effect on the T levels of male rats. However, centrifugation at 4 *g* for 4 h resulted in reduced T levels, suggesting that duration of exposure is important. The hypergravity experienced at lift-off and re-entry is short term; therefore, it is unlikely that the reduced T in space flight rats was due to acute hypergravity. Chronic centrifugation at 2 *g* had no effect on testes weight, T, or spermatogenesis compared with control rats, but 2.3 *g* reduced T during the initial 3 days, and 4.1 *g* resulted in continuous suppression of T throughout 52 days of centrifugation. These results suggest that over time, rats are capable of adapting to gravitational forces of >2.3 *g* but not to 4.1 *g*. Using centrifugation to validate the space flight findings has been difficult because of the pulsatile secretion of LH and elevated LH in response to disturbing stimuli such as the stopping of the centrifuge. Ortiz et al., avoiding the bias of daily periodicity by measuring T in 24-h urine samples, found a transient increase in urinary T in male rats centrifuged at 2 *g*. This finding contrasts with those of other reports of reduced T in response to centrifugation. Stress rather than hypergravity may have been responsible for elevated T, because elevated T has been reported as a response to acute stress. Clearly more studies are needed, but the results of these studies suggest that hypergravity altered male reproductive hormone levels and that the response of the HPG axis depends on duration and *g* force.

Effects of Space Flight on Female Reproduction

The number of female astronauts is growing, and the presence of women on long-duration spaceflights and at the International Space Station is also expected, yet relatively little is known about the effect of space flight on female reproduction. Environmental conditions can exert a number of adverse effects on the ovaries. To our knowledge, no researchers have examined the effect of the space environment on the ovaries of nonpregnant females. However, examination of the ovaries of postpartum rats flown in space during Days 9–20 of gestation showed no effect on ovarian weight or number of preovulatory or atretic follicles. According to Ying et al., the mechanism mediating postpartum ovulation is the same as that in nonpregnant rats, suggesting that space flight does not affect the ovaries. Reproductive changes do not always occur at the ovaries but may occur at the hypothalamus or pituitary. The HPG axis regulates the ovulatory cycle, which is highly susceptible to environmental factors. For example, stress can lead to ovulation failure. Thus, the amount of stress experienced during space flight might be correlated with negative effects on the ovulatory cycle, but this possibility has not been explored. No data on women have been collected because female astronauts suppress their menstrual cycles during space flight. Animal studies of altered gravity examining ovulatory cycles are also scarce. In postpartum space flight rats, ovulation was suggested to be suppressed based on findings of reduced pituitary LH; however, measurement of plasma LH indicated no change. Plasma FSH was also elevated; however, there was no effect on pituitary FSH. These unequivocal results are likely due to the pulsatile secretion of LH and FSH. More meaningful results would require sequential blood sampling for LH and FSH. Serova and Denisova reported that rats mated in hypogravity during space flight, indicating that female rats ovulated and cycled normally; however, no births resulted. Postflight laparotomies suggested that the fetuses were resorbed, but there is no clear indication of whether conception or implantation can proceed in the space environment. No systematic studies have examined the effect of space flight on early pregnancy. However, several space flight missions have included rat dams for varying durations between Days 9 and 19 of the 22-day gestation period. The first space flight mission carrying pregnant mammals reported a reduction in pregnancy weight gain, prolonged parturition, lower birth weights, and increased perinatal mortality. Increased neonate mortality persisted into the F_2 generation. Diet, housing, and other environmental conditions peculiar to this flight may have contributed to the adverse pregnancy outcomes. Another factor may have been stress; prenatal stress is associated with increased fetal mortality, abnormal nursing behaviour, and increased postnatal mortality. However, subsequent space flight studies have yielded no evidence that pregnant rats are stressed during space flight.

Despite speculation that headward fluid shifts, cardiovascular deconditioning, bone demineralization, and decreased red cells associated with hypogravity may affect the ability of rat dams to sustain their pregnancy, the results of space flight studies indicate that pregnant rat dams are able to successfully direct physiological resources to support fetal development in the space environment. In females, appropriate ratios of estrogen and progesterone are required for the establishment, maintenance, and termination of pregnancy.

In view of the changes in male reproductive hormones following space flight, there is concern that space flight also may affect female reproductive hormones. However, Burden et al., reported no change in estrogen and progesterone levels in pregnant space flight rats.

Space flight rat dams did experience an increased number of contractions during parturition due to reduced connexin 43, the major gap junction protein in the myometrium that plays a role in the synchronization and coordination of contractions. Although contractions in space flight rat dams were increased, there were no effects on fetal wastage, birth weight, litter size, or maternal care of the neonates. Although progress has been made in the area of late pregnancy and neonate development, more studies are needed on fertility, conception, and early pregnancy in space. Ground-based models are well suited to address these existing gaps in our current knowledge.

Effects of Simulated Hypogravity on Female Reproduction

Ground-based models have not been used frequently to study the effects of hypogravity on female reproduction. Women exposed to 17 days of 6° head-down tilt bed rest showed no changes in menstrual cycle length. However, the duration of the study was too short to draw definitive conclusions. Rock and Fortney reported that women exposed to bed rest exhibited luteal phase deficiency, a condition that is related to HPG axis dysfunction. However, this study lacked controls. Clearly, more studies are needed to answer basic questions about female reproductive processes and to address concerns that fluid shifts associated with hypogravity may lead to retrograde menstruation or endometriosis.

To our knowledge, no researchers have used HLS to investigate hypogravity effects on reproductive processes in nonpregnant female rats, although HLS is ideally suited to answer pertinent questions. There is no concern about the translocation of reproductive organs in female rats because the ovaries are normally situated within the abdominal area. In addition, monitoring of the ovulatory cycles of HLS female rats is a simple technique that can provide important information on ovulation, reproductive capability, HPG, and ovarian function in nonpregnant females. A few HLS studies have been conducted on pregnant rats. HLS of rats during early pregnancy reduces implantation. The absence of elevated plasma or adrenal corticosterone in HLS pregnant rats suggests reproductive effects were due to simulated hypogravity exposure rather than stress. Similar findings derived from the space flight experiments raise questions as to whether early pregnancy can be sustained in hypogravity.

Effects of Hypergravity on Female Reproduction

Hypergravity exposure of nonpregnant rats produced no differences in mating ability or gestation, but fewer pregnancies resulted. Hypergravity at moderate levels of 1.46 and 2.28 *g* disrupted the ovulatory cycle by inducing prolonged diestrus in rats, suggesting that the reduction in the number of pregnancies observed in centrifuged rats is due to changes in the ovulatory cycles. Pregnancy did not occur in mice chronically centrifuged at 3.5 *g*. Even centrifugation at relatively low speeds of 1.41–1.47 *g* resulted in an increased tendency to abort, especially during early pregnancy. Serova observed that centrifugation of rat dams at 2 *g* during early pregnancy resulted in a 33% incidence of spontaneous abortion, whereas in rats centrifuged during midpregnancy there were no cases of spontaneous abortion. These findings and the results of space flight studies suggest that exposure to gravitational forces different from Earth's 1 *g* may interfere with early pregnancy events such as implantation.

Increased mortality has been reported in neonatal rats exposed to hypergravity, even with intermittent exposure to centrifugation during the periparturition period. Oyama et al. reported that pregnant rat dams centrifuged at 2.16 or 3.14 *g* had reduced numbers of survivors within their litters compared with 1-*g* controls. Furthermore, neonatal mortality in this study increased with increasing *g* load. Survival and development of neonates in hypergravity is partly dependent upon normal maternal behaviour and mother-offspring interactions. Decreased plasma prolactin in rat dams centrifuged at 2.16 or 3.14 *g* has been suggested as a mechanism responsible for decreased neonatal survival because prolactin plays a key role in the expression of maternal behaviour in rats. Because maternal care may be influenced by previous experience, Ronca et al. compared primigravid and bigravid rats exposed to 1.5 *g* centrifugation at Gestation Days 11–22. Pup survival was 82% for primigravid dams, 94% for bigravid dams, and 99% for 1-*g* controls, suggesting that previous maternal experience is protective against neonatal loss. These study results indicate that altered gravity affects reproductive physiology during pregnancy and lactation and affects maternal behaviour and mother-offspring interactions. Further ground-based studies are needed to provide a more detailed understanding of the issues of pregnancy, birth, and rearing of offspring in space.

Effects of Altered Gravity on Fertility

Ultimately, successful reproduction is the production of viable progeny. In only one study was mating capability in the space environment examined. Rats were described as having mated successfully in hypogravity, although no viable progeny were produced. It is unclear from the study design whether reproductive effects occurred in the males, the females, or both.

Male space flight rats mated 5 days after flight to nonflight female rats bred successfully, but their offspring showed growth retardation and more frequent abnormalities, such as edema, hemorrhages, hydrocephaly, ectopic kidneys, and enlargement of the bladder.

Male progeny also showed reduced epididymis weight at 30 days of age, although this reduction did not persist into adulthood (100 days of age). However, male rats mated 2.5–3 mo after space flight produced healthy, viable offspring. Santy et al. suggested that space flight produced abnormalities in mature sperm but had no effect on early stage sperm. Anecdotal evidence indicates that male astronauts produce healthy offspring following space flight. To our knowledge, fertility following recovery from HLS has not been explored. The effect of hypergravity on fertility was investigated by exposing jointly caged male and female rats to chronic centrifugation. Results indicated some reduction in pregnancy rate at 2 *g* and no successful pregnancies at 4.7 *g*, an extremely high *g* load. Megory et al. observed prolonged diestrus in centrifuged rats. Changes in the ovulatory cycle can compromise fertility. Reports of reduced T in centrifuged male rats can also compromise fertility because T is important for maintaining reproductive organ function and spermatogenesis. Whether long-term exposure to altered gravity can lead to subfertility or even infertility and whether effects are temporary or permanent are important questions that must be answered before long-term habitation of space is attempted.

Conclusions

As the International Space Station moves us closer to the reality of space colonization, it becomes increasingly important to understand the effects of altered gravity on mammalian reproductive physiology and function. Based on the existing data, there is evidence for hypo- and hypergravity-induced changes in male and female reproductive processes. However, additional research is needed to fill in the gaps in our current knowledge. Although space flight experiments are critical, ground-based studies are important for the correct interpretation of space flight findings because space flight studies are often encumbered by flight restrictions and effects of environmental factors unrelated to hypogravity (i.e., radiation, vibration, noise). Research incorporating ground-based models is yielding results quite similar to those obtained from space flight. These similarities confirm the appropriateness of HLS, bed rest, and centrifugation for studying reproductive physiological responses and adaptation to altered gravity. Together, these unique tools are yielding new insights into the gravitational biology of reproduction in mammals.

Deconstructing Mammalian Reproduction: Using Knockouts to Define Fertility Pathways

Reproduction is the *sine qua non* for the propagation of species and continuation of life. It is a complex biological process that is regulated by multiple factors during the reproductive life of an organism. Over the past decade, the molecular mechanisms regulating reproduction in mammals have been rapidly unraveled by the study of a vast number of mouse gene knockouts with impaired fertility. The use of reverse genetics to generate null mutants in mice through targeted disruption of specific genes has enabled researchers to identify essential regulators of spermatogenesis and oogenesis *in vivo* and model human disorders affecting reproduction. This review focuses on the merits, utility, and the variations of the knockout technology in studies of reproduction in mammals.

Living forms are defined by their ability to reproduce. Although different organisms produce progeny using diverse methods (asexual and sexual reproduction, parthenogenesis, etc.), the basic principles of reproduction, both at the molecular and the organismal levels, have remained the same since the dawn of life: formation of gametes to facilitate the transfer of genetic material and epigenetic information to the next generation. Indeed, in several model organisms, orthologous genes have been identified that perform similar roles in reproductive physiology, thereby highlighting a conservation of this complex biological process.

Historically, two major obstacles have impeded the study of reproductive biology in mammals (reviewed by Staub 2001). First, germ cells in mammals develop in a microenvironment of supporting stromal cells of somatic origin that interacts with the former through autocrine/paracrine mechanisms as well as direct cell-to-cell interactions. Devoid of this support, isolated germ cells in culture fail to survive and maintain their characteristics, making it highly difficult to study molecular interactions and pathways that are unique to these cells.

The second, but related, hindrance has been the inability to develop a proper cellular context *in vitro* that would allow undifferentiated germ cells to differentiate into specialized gametes that retain their ability to fertilize and produce healthy offspring. Recently, interest has exploded in the field of *in vitro* germ cell differentiation due to exciting findings by several investigators who have been able to generate meiosis-competent germ cells from undifferentiated embryonic stem cells. Although these gametes have been shown to fertilize oocytes and produce blastocysts, no data are available yet to suggest that these can progress to produce live offspring. In addition, Feng and coworkers (2002) have generated an undifferentiated spermatogonial cell line immortalized by the ectopic expression of the catalytic component of telomerase. The cell line was also shown to differentiate *in vitro* to morphologically similar haploid round spermatids but the authors did not investigate whether these haploid germ cells were competent to initiate fertilization and zygotic development when injected into oocytes. So, although the future looks promising, more careful analysis of the physiological state of these *in vitro*-formed gametes, including the identification of parent-of-origin-specific imprints, would be critical for enhancing the utility of these techniques.

Our understanding of the molecular mechanisms of mammalian reproduction therefore has been largely dependent on loss-of-function mutagenesis in mice. Ever since researchers developed techniques to culture mouse embryonic cells *in vitro* and methods were developed to inject cultured embryonic pluripotential cells into blastocysts to produce germ-line chimeras, the field of mouse transgenic technology has undergone a veritable explosion. Around the same time, researchers were successfully targeting exogenous DNA into endogenous loci based on limited amounts of homology between the two DNA sequences in cultured mammalian cells. Simultaneously, Thomas and colleagues had developed a highly efficient method of targeted mutagenesis of cultured embryonic stem cells and other mammalian cells by using homologous recombination between endogenous genes and exogenously introduced DNA (targeting vector), a technique that was shown to be highly specific by Doetschman and coworkers. Our present-day knockout technology, a cornerstone of functional genomic studies, got its final form when these two technologies were harnessed together to generate the first targeted knockout mouse model through homologous recombination in embryonic stem cells.

A knockout mouse model is typically designed to produce a null mutation in the target gene and thus uncover an essential function. In dissecting gene functions, a hypomorphic mutation or the generation of an allelic series can at times be beneficial to researchers testing for genetic interactions between candidate genes. However, a null mutation is preferred in identifying non-redundant and essential functions of genes, as exemplified by the generation of approximately 300 mouse knockouts that manifest defects in reproduction. Among the earliest mouse knockouts generated that were found to affect different aspects of reproduction were *Inha* (inhibin), *Lif* (leukemia inhibitory factor), and estrogen receptor. While LIF-deficient females had defects in implantation, knockout males and females of inhibin were infertile and had gonadal tumours. Female mice null for estrogen receptor were infertile, and males had reduced fertility. Over the last decade, several hundred mouse knockouts have been created that affect different stages of reproduction, including primordial germ cell (PGC) formation and migration into the gonads, gonad development and sex determination, postnatal gametogenesis, sperm–egg binding and fertilization, and early embryonic development.

In this review, we focus, in a stage-specific manner, on specific knockouts that have proved pivotal in our understanding of the biology of reproduction, with special emphasis on a few *in vivo* models that have clarified or refuted conflicting findings from *in vitro* experiments. We then briefly describe the principle of conditional mutagenesis and highlight its use in reproductive studies. Finally we discuss some of the current deficits of these technologies that have tempered the enthusiasm of researchers embracing them as their primary tools in functional genomics.

Germ-line Specification: What Seals the Fate?

In most model organisms studied so far, the germ line is specified in embryos by a maternally derived source of egg cytoplasm referred to as the germ plasm. In mammals, however, germ-cell determination appears to be regulated within the classical developmental paradigm of cell signalling and cell adhesion.

In mice, PGCs are the founder cells of the germ line that are allocated from a pool of proximal epiblast cells at around embryonic day 7.2. By using mice that were deficient for bone morphogenetic protein (BMP), member 4 (*Bmp4* $^{-/-}$ mice), Lawson and colleagues (1999) provided clear evidence that germ-cell competence is induced in a subset of proximal epiblast cells in response to BMP4 secreted from the extra-embryonic ectoderm. In *Bmp4* $^{-/-}$ mice PGCs were completely absent, and in *Bmp4* heterozygotes the PGC number was either greatly reduced or completely absent, suggesting that the effect of BMP4 is dose-dependent. Deficiency of the related transforming growth factor ß (TGFß) superfamily members, BMP8b and BMP2, led to a very similar phenotype of reduced or absent PGCs on a mixed genetic background. Interestingly, mice null for SMAD5 or SMAD1, the intracellular effectors of BMP signalling, have pronounced defects in PGC formation as well. More recently, knockout mice carrying a targeted mutation in the type I BMP receptor, activin receptor-like kinase 2 (ALK2), were found to have a complete absence of PGCs, further highlighting the importance of this signalling pathway in germ-cell specification.

While these studies suggest that an intact BMP signalling system in the embryonic stages is necessary for PGC fate specification, what is not clear is whether such a signalling mechanism is actually sufficient to induce PGC fate. What are the primary targets of BMP4 and BMP8b signalling in the epiblast? Do they signal directly to the PGC precursors? Is there a master regulator of PGC fate? These are some of the questions that remain unsolved. However, a recent study, utilizing a knockout in the mouse zinc-finger-containing transcriptional repressor PRDM1 (PR domain containing 1, with zinc-finger domain), has uncovered an essential role for this protein in PGC specification. Embryos lacking the PRDM1 protein have a complete absence of PGCs while heterozygotes have a drastic reduction in PGC number. Since *Prdm1* transcripts are expressed in PGC precursors as early as E 7.0, this is the first transcriptional factor to be identified that plays an essential role in PGC specification. So, is PRDM1 the master regulator that seals the fate of PGC precursors into entering a germ-cell lineage? The answer to that question awaits further studies, which would address key questions regarding the sufficiency of PRDM1 in specifying a germ-cell fate upon non-committed epiblast cells and whether its action is cell-autonomous in these precursors.

Studying gene function through knockouts has proved highly significant in the field of reproduction not only because of the uncovering of new genes and pathways but also because these *in vivo* studies on many occasions served to prove or refute several models that were based on either cell-culture-based assays or on evidence from forward-genetics approaches. One instance is noteworthy: by analyzing the expression pattern of genes from solitary nascent PGCs, Saitou and colleagues (2002) proposed a mechanism for acquisition of germ-cell competence and specification in PGC precursors in the epiblast which was based on essential roles for two highly expressed and unique cDNAs (named *fragilis* and *stella*) in PGC fate determination. However, when the *stella* gene was knocked out in mice, there appeared to be no defects in the specification, maintenance or proliferation of PGCs; instead, knockout males were fertile although null females failed to sire normal numbers of offspring as the gene fulfilled an essential role in the pre-implantation development of embryos as a maternally derived factor. In this context it would be interesting to know the phenotype of mice with a null mutation in *fragilis*.

PGC Proliferation, Migration and Survival

Subsequent to their specification from precursor cells, PGCs undergo several rounds of proliferation till E 13.5, around when they enter cell-cycle arrest in G_0 within the male gonad or meiotic prophase I in the female gonad. In between, they migrate to the gonads by E 10.5 after travelling through the hindgut and the dorsal mesentery in response to secreted cues from the environment.

Through knockout approaches, researchers have identified several new genes that play essential roles in some or all of these processes.

Targeted disruption of a peptidylprolyl isomerase-encoding gene, *Pin1*, uncovered an unexpected role for this gene in the regulation of the cell cycle and PGC proliferation. On an isogenic C57BL/6J background, mice null for *Pin1* were found to have reduced PGC numbers during the mid-to-late migratory phases compared with wild-type mice; by analysing BrdU labelling, this reduction seemed to be due to a delay in progression through the cell cycle rather than due to a cell-cycle arrest or enhanced apoptosis in these cells. Specific genes regulate the survival of the migratory germ cells: the POU domain transcription factor, *Oct4*, is a critical determinant of pluripotency in the inner cell mass that causes embryonic death when mutated.

To circumvent the early embryonic death and to examine the effect of deletion of *Oct4* from PGCs, investigators have used *TNAP* ^{Cre} (tissue non-specific alkaline phosphatase – Cre), which encodes the Cre recombinase only in the PGC population prior to E 10.5. Surprisingly, loss of OCT4 protein from PGCs leads not to differentiation but to enhanced apoptosis in these cells prior to their colonization of the gonads. *Nanos3*, a mouse homolog of the *Drosophila nanos* gene, represents another interesting knockout model in which the mouse homolog was found to perform an evolutionarily conserved role in flies and mammals.

In *Nanos3* $^{-/-}$ mice, PGC proliferation and/or survival during migration through the hindgut was severely disrupted, leading to sterility in both sexes. In further examples of evolutionary conservation of these mechanisms knockout mice deficient in the chemokine stromal cell-derived factor-1 (Sdf1) and its cognate receptor CXCR4 (chemokine C-X-C motif receptor 4) show defective colonization of the gonads by PGCs. PGCs that successfully navigate to the genital ridges enter cell-cycle arrest, although at different stages in the two sexes, as mentioned above. Heterozygosity for the transcription factor zinc-finger protein 148 (*Zfp148*) has been shown to increase apoptosis and cause complete loss of germ cells from the male gonad at the E 13.5 stage, an effect that is probably due to improper regulation of the p53 protein in these cells. Thus *Zfp148* is one more addition to the growing list of genes that seem to affect PGC specification and maintenance in a dosage-dependent manner in mice.

Sex Determination

Acquisition of reproductive competence in mammals involves the development of both germ cells and the supporting somatic cells that together constitute the gonad. The development of either the testis or the ovary from a bipotential gonad is a complex developmental process that is influenced by several transcriptional pathways. The first gene that was identified to play an essential role in determining the sex of the gonad was the Y-chromosome-encoded putative transcription factor *Sry*. In mice the *Sry* gene product has been shown to be both necessary and sufficient to impose a testis fate on the bipotential gonad. Although a recent report described a role for SRY in pre-mRNA splicing, downstream targets of this master sex-determining gene remain elusive.

Our understanding of sex determination and gonadal organogenesis therefore have been dependent on targeted mutagenesis of factors that have been implicated to regulate *Sry* or to affect gonad formation in other model organisms or through spontaneous mutations leading to sex-reversal phenotypes in humans and mice (reviewed in Brennan & Capel 2004). Among these, the X-linked gene encoding the orphan nuclear receptor DAX1 (NR0B1, nuclear receptor subfamily 0, group B, member 1; Mouse Genome Informatics) was initially believed to cause XY sex reversal in mice when over-expressed in a *Sry* hypomorphic mutant background. Since over-expression of DAX1 led to the development of female reproductive structures from a bipotential gonad, albeit on a sensitized background, it was assumed that this gene was critical for ovarian development. Intriguingly, targeted deletion of *Dax1*, employing a conditional allele, does not lead to failure of ovarian development; on the contrary, postnatal spermatogenesis is impaired while null females are fully fertile.

Furthermore, when the *Dax1* $^{+/-}$ females were mated to males with the hypomorphic *Sry* allele, all XY offspring were sex-reversed and their external and internal genitalia were completely indistinguishable from their wild-type XX siblings, suggesting that this gene interacts with *Sry* to determine gonadal sex in mice and is required not for ovarian development but for spermatogenesis, probably through its action in Sertoli cells. It is critical to note, however, that the role of DAX1 is highly dependent on the genetic background of the mice under investigation. A recent study by Eicher and colleagues demonstrated that although DAX1 function is essential on the C57BL/6 (B6) background for testis development in XY mice, it is completely dispensable on the DBA/2J background. In B6 *Dax1* $^{-}$/Y mice, there was a failure of up-regulation of *Sox9*, which is another important gene in the sex-determination pathway, even though *Sry* expression initiated normally.

The homeobox gene *Sox9* is the earliest marker of pre-Sertoli cells following *Sry* expression. Through the use of a conditional knockout model SOX9 has been shown to be essential for normal testis differentiation and sex-cord formation in the testis. SOX9 regulates the differentiation of Sertoli cells and the production of male-specific markers *Amh* (anti-Müllerian hormone) and P450scc (cytochrome P450 side-chain cleavage), both of which are decreased in homozygous knockout mice. Indeed, this finding is a corroboration of earlier studies in which targeted disruption of the SOX9-binding site in the mouse *Amh* promoter caused a failure of transcription of *Amh* that resulted in non-regression of the Müllerian duct structures in XY mice and the formation of pseudo-hermaphrodites, confirming that SOX9-mediated up-regulation of AMH is a critical and specific step that is essential in the male sex-determination pathway.

Precisely how AMH mediates regression of Müllerian duct structures *in vivo* has been characterized by the generation of a conditional knockout for one of the type I BMP receptors, *Bmpr1a*; male mice lacking these receptors in mesenchymal cells of the Müllerian duct fail to respond to AMH and have persistent uteri and oviducts.

Two isoforms of the WT1 (Wilms' Tumor 1) protein, produced by alternative splicing, that differ in the presence or absence of 3 amino acids (KTS; the +KTS and –KTS forms) play unique roles in gonad development and sex determination.

By introducing point mutations in the gene to create mutants expressing either one of the two isoforms, deficiency of the –KTS form was found to cause gonadal agenesis in both sexes whereas a loss of the +KTS form uncovered a critical role for this isoform in male sex determination as both XX and XY homozygotes developed along the female pathway. Since *Sry* expression was reduced in the KTS-ablated WT1 homozygotes, it can be inferred that WT1 + KTS is important for *Sry*-mediated XY gene program establishment in the bipotential gonad; whether this down-regulation is a direct result of WT1 + KTS loss or as a result of decreased numbers of *Sry*-expressing pre-Sertoli cells awaits further experiments.

Another gene involved in ovary development is *Wnt4*, which is required for Müllerian duct formation in both sexes prior to sex determination; female mice lacking *Wnt4* exhibit partial female–male sex reversal.

Since *Fst*-null ovaries have a similar defect and since *Fst* expression is absent in *Wnt4*-null XX gonads, *Fst*, operating downstream of *Wnt4*, has been implicated in the suppression of BMP and/or activin signalling to suppress the male-specific coelomic vessel.

Molecular Control of Pre-antral Folliculogenesis

Folliculogenesis in the mammalian ovary is one of the most intricately regulated developmental processes in biology. Not only does normal folliculogenesis entail the complex interplay of both long-range endocrine cues and short-range paracrine and autocrine factors, the development of viable offspring from fertilization depends, to a large extent, on the structural and functional integrity of the mature oocyte.

Physiologically, it is useful to visualize ovarian follicles as divided into two broad groups, each being characterized by the presence or absence of a morphological antrum (cavity). Follicles prior to the formation of an antrum are independent of the pituitary gonadotropins for growth and survival; these pre-antral follicles undergo an extensive growth phase once they have been recruited into the active pool although they remain arrested in a prolonged diplotene stage of meiosis throughout this phase. In contrast, around the period of antrum formation, oocytes regain meiotic competence and oocytes in early antral follicles depend on pituitary follicle stimulating hormone (FSH) for survival and growth.

The formation of primordial follicles in the newborn mouse ovary is preceded by the development of germ-cell clusters in the embryonic ovary in which the dividing germ cells undergo incomplete cytokinesis to remain connected to each other through intercellular bridges and their subsequent breakdown perinatally. The search for active players that regulate the formation of primordial follicles in mammals has resulted in the discovery of a bHLH (basic helix-loop-helix) transcription factor called FIG (factor in the germ line; *Figla*) present specifically in the female germ cells.

Mice carrying a targeted disruption of the *Figla* gene (*Figla* $^{-/-}$) had comparable number of germ cells in their ovaries with respect to wild-type ovaries at the E 18–19 stage. However, by postnatal day 1 (PD 1) the oocytes in the *Figla* $^{-/-}$ ovaries failed to grow in size and began to disappear; by PD 2 the null ovaries were completely devoid of primordial follicles. The *Figla* $^{-/-}$ oocytes do progress through meiosis I, but fewer cells survive beyond the diplotene stage, suggesting that FIG regulates the expression of survival factors/growth factors at this stage.

While *Figla* represents an oocyte-derived factor critical for the formation of primordial follicles, data from another mutant provide evidence that primordial follicle formation is regulated by the somatic components as well. Mutants lacking the *Wnt4* gene product die shortly after birth due to defects in kidney development; however, at day 1 the mutant ovaries have a dramatic reduction in the number of meiotic-stage oocytes and most oocytes degenerate shortly thereafter. A similar finding has emerged from mice with follistatin-deficient ovaries, highlighting the functional interactions between these two proteins.

Shortly after birth a subset of primordial follicles is selected from a resting pool of quiescent follicles to undergo the follicular activation and growth. Follicular activation is characterized by oocyte growth and the transition of squamous to cuboidal granulosa cells, which signals the formation of a primary follicle followed by granulosa cell proliferation. Work from our group and others has recently identified several proteins believed to be critical at this step. A homeobox transcription factor, NOBOX, which is specific to the oocyte in the newborn ovary, is essential in this process. In *Nobox* $^{-/-}$ ovaries, primordial follicles do not develop into primary follicles and beyond; instead, by PD 14, the majority of the primordial follicles are lost leading to ovarian failure and infertility. Future studies will illustrate the target genes for this transcription factor and uncover other oocyte-specific genes that play critical roles in this process.

Primordial follicles are recruited in cohorts for growth, differentiation, and maturation through a series of highly regulated processes. However, primordial follicles in *Foxo3a* $^{-/-}$ ovaries undergo global activation so that by PD 14 the null ovaries do not have any quiescent primordial follicles but have an almost 2-fold increase in the number of primary and secondary follicles as compared with wild-type ovaries. Significantly, the null ovaries also have an almost 4-fold increase in atretic follicles and an approx. 14-fold increase in the amount of zona pellucida remnants, suggesting that the majority of the growing follicles undergo atresia following the formation of the zona pellucida by the oocyte. Not surprisingly, therefore, this sequence of global activation and atresia leads to a premature depletion of the pool of primordial follicles and secondary infertility by 15 weeks of age.

Mutations in the human forkhead transcription factor *FOXL2* cause an autosomal dominant condition referred to as Blepharophimosis/ptosis/epicanthus inversus syndrome (BPES), associated with premature ovarian failure. Mice with a targeted disruption in the murine ortholog (*Foxl2* $^{-/-}$ mice) of the human gene are infertile and provide a mechanism to explain the etiology of BPES. In *Foxl2* $^{-/-}$ mice granulosa cells in the mutant ovaries do not make the squamouscuboidal transition and consequently the mutants lack any secondary follicles and are infertile. Another granulosacell derived factor that regulates the rate of follicle recruitment, in a subtler manner, is the anti-Müllerian hormone (AMH), which, when secreted from the granulosa cells of secondary follicles, acts in a paracrine fashion within the follicle to inhibit the recruitment of primordial follicles. In *Amh* $^{-/-}$ females there is a derepression of the follicular activation process causing mutant ovaries to be depleted of primordial follicles earlier than in wild-type ovaries.

Growth differentiation factor 9 (*Gdf9*) is an oocyte-derived mitogenic paracrine factor that plays a critical role in folliculogenesis by regulating the somatic components of the follicle. *Gdf9* encodes for a member of the TGFß super family of secreted proteins that play important roles in cell proliferation and differentiation in mammals. Members of this family, such as GDF9, BMP15, AMH, and activins, exquisitely regulate early stages of folliculogenesis. In mice, a null mutation in *Gdf9* causes complete infertility in female mice. Histopathological examination of null ovaries demonstrate a total lack of normal appearing secondary follicles indicating that GDF9 secreted from oocytes acts upon the granulosa cell population to augment the proliferation/differentiation of these follicular supporting cells. Thus, GDF9 is essential for the primary-to-secondary follicle transition in mice. The oocyte that has already entered the growth phase following activation continues to grow in these follicles; however, devoid of the close support from the granulosa cell mass there is atresia of both the oocyte and eventually the somatic cells.

Another member of the TGFß super family of proteins that appears to play an important role in pre-antral stages of folliculogenesis is BMP15. BMP15 is secreted from oocytes and regulates the proliferation and differentiation of the granulosa cell population. Analysis of a null mutation in *Bmp15* generated in our laboratory revealed that *Bmp15* $^{-/-}$ mice are subfertile and display defects in ovulation and fertilization. However, studies in sheep have shown that BMP15 may also play important roles at the primary to secondary follicle transition or, alternatively, the mutant BMP15 may antagonize the functions of GDF9.

3

Spermatogenesis

In adult mammals, spermatogenesis is one of the most highly productive biological processes; one study done on human testis specimens obtained at autopsy estimated that the daily sperm production ranges between 3.5 and 7.5 million sperm/g of testis parenchyma. This tremendous capacity to produce spermatozoa throughout the life of an individual rests solely on the self-renewing capacity of spermatogonial stem cells, which constitute a small group of pluripotent cells present in the mature testis. Concomitant with the identification of molecular markers that identify the spermatogonial stem-cell pool has come new knowledge on the regulation, survival and proliferation of these rare populations of adult stem cells *in vivo* through knockout studies. Here we briefly review some of these models that have illustrated the utility of the targeted knockout approach.

Among the first mutations that were identified to have a defect in spermatogonia were the *W* and *Sl* mutations (Russell *et al.* 1979) that affect the Kit receptor, which is present on differentiating spermatogonia and its ligand, the Stem cell factor, which is produced in the Sertoli cells. However, mutations in the Kit signalling system affect only the differentiated spermatogonia as the receptor Kit is not expressed on earlier spermatogonial populations. Recently, glial cell-derived neurotrophic factor (GDNF), another Sertoli cell-derived factor distantly related to the TGFß family of growth factors, was shown, in a dose-dependent manner, to control the cell-fate decision of undifferentiated spermatogonia; reduced levels of GDNF caused spermatogonial differentiation and depletion of the stem-cell pool while high levels *in vivo* caused a differentiation block and increase in the numbers of spermatogonia.

Germ-cell factors that play equally important roles in the proliferation of undifferentiated spermatogonia have also been identified through mouse knockouts; prominent has been the identification of a zinc-finger-containing transcriptional repressor, *Plzf*, that is essential for stem-cell self-renewal in the adult testis. Although the precise genes that are regulated by this transcription factor are not yet known, future studies utilizing transcriptome analysis of spermatogonia similar to the approach used by Costoya and colleagues (2004) should lead to further insights into the molecular mechanisms of germ-line stem-cell biology.

Knockouts have been created that identify essential genes at almost all steps of spermatogenesis; literally hundreds of mutant mice have been characterized with arrest at various stages such as a spermatogonial differentiation block, pachytene stage meiotic arrest, and defects in spermiogenesis and sperm function.

Several genes that are expressed in post-meiotic cells have been found to be essential for spermiogenesis in a dose-sensitive manner. Notable among these are mutations in the genes encoding protamines 1 and 2 (*Prm1* and *Prm2*), Kelch-like 10 (*Klhl10*) and sperm antigen 16 (*Spag16*), haploinsufficiency of which are incompatible with male fertility. Male mice that are high-percentage chimeras of these genes are unable to transmit the mutant alleles through the germ line due to either severe morphological defects (as in the *Prm1* or *Prm2* mice) or severe reduction of sperm count and defects in sperm morphology and function (*Klhl10* or *Spag16*). During spermiogenesis, chromatin packaging in the sperm head requires the replacement of histones by the transition proteins and protamines. Similar to protamines, deficiency of both transition proteins 1 and 2 also leads to severe defects in sperm morphology and infertility (Zhao *et al.* 2004), thereby highlighting the importance of proper chromatin remodeling during spermatogenesis. Defects in sperm motility also lead to infertility; analysis of knockouts of *CatSper1* and *CatSper2*, which are cationic channels present on the sperm membrane, a sperm-specific isoform of a glycolytic enzyme, *Gapdhs*, and of structural components of the axoneme (*Spag6*, *Tekt2*) have revealed interesting models of asthenozoospermia that lead to infertility in male mice. Finally, knockouts have uncovered genes that play essential functions in the final step of sperm function, sperm–egg binding. The plethora of single-knockout models that impair spermatogenesis at so many discrete steps is in addition to numerous models that cause a hitherto undefined or non-specific block in spermatogenesis and as a whole serve to illustrate both the complexity of this process and the huge number of potential targets for development of contraceptives in humans.

Endocrine, Paracrine and Autocrine Regulation of Gametogenesis

Germ cells are dependent on the somatic cells of the gonad (Sertoli and Leydig cells in the testis; granulosa and theca cells in the ovary) for development, which are in turn regulated by gonadotropins secreted by the pituitary, namely FSH and luteinizing hormone (LH). Deficiency of the pituitary gonadotropins leads to hypogonadism in both sexes, as in the *hpg* (hypogonadal) mouse, which carries a naturally occurring deletion in the gonadotropin-releasing hormone (GnRH) gene and consequently are infertile with reduction in the size of the gonads and accessory sex organs as well as a block in gametogenesis. Since *hpg* mice lack both LH and FSH, the relative contributions of each of these gonadotropin hormones to the phenotypic spectrum was not clear until our group and others created knockouts of four different genes in mice: *Fshb* (FSHß subunit), *Fshr* (FSH receptor), *Lhb* (LHß subunit) and *Lhr* (LH receptor).

Analysing the effects of these single knockouts revealed that FSH and LH had non-redundant roles in both sexes. In males, loss of FSH signalling resulted in smaller testis size, reduced spermatogenesis, and impaired sperm motility although the males were fertile. In contrast, females deficient for a functional FSH signalling pathway are sterile with thin uteri and hypoplastic ovaries due to a block in folliculogenesis prior to antrum formation. An intact LH signalling pathway, however, is essential in both sexes as both males and females deficient for either LHß subunit or its cognate receptor develop hypogonadism and infertility. Furthermore, folliculogenesis is blocked in LH-deficient ovaries at an early antral stage, suggesting that the progression from secondary to early antral follicle stage is dependent on FSH while LH is required for further development. Similarly in males, spermatogenesis is blocked at the round spermatid stage, revealing an absolute requirement for only LH in further maturation of the haploid gametic cells.

Sex hormones (estrogen and progesterone in females, testosterone in males) have been long believed to play essential roles in the development, differentiation and proper function of female and male reproductive systems. Targeted knockouts of their respective receptors have revealed precisely how these steroid hormones function and have uncovered hitherto unsuspected roles for estrogen signalling in males and testosterone in females. Estrogen acts through two nuclear hormone receptors, ER and ERß; whereas deletion of ERß leads to subfertility due to defects in granulosa cell development, disruption of ER results in female infertility due to a complete lack of ovulation with the null ovaries displaying hemorrhagic cysts.

Interestingly, ER-knockout males are infertile as well due to a defect in fluid reabsorption in the epididymis that leads to degeneration of seminiferous tubules and disruption of spermatogenesis, thereby identifying a male-specific role for estrogen. Progesterone, which also acts through two receptor isoforms (PR-A and PR-B), was believed to be important for the establishment and maintenance of the pregnant state; however, deletion of the progesterone receptor in females led to infertility due to defects in ovulation with associated uterine inflammation and disruption of mammary gland morphogenesis. Identifying which of these effects were mediated by the two receptor isoforms required some subtle manipulations because both isoforms were generated from the same gene from alternative promoters and translation initiation sites. In elegant experiments, Mulac-Jericevic and colleagues mutated the two AUG initiation codons in embryonic stem cells and generated PR-A and PR-B knockouts. Analysis of these mutants revealed essential roles for PR-A in ovulation and uterine physiology while PR-B was important for mammary gland development.

Studies from the spontaneous *tfm* (testicular feminized) mutant and from androgen receptor-knockout mice have shown androgen signalling to be essential for the descent of testes, development of male accessory sex organs, and spermatogenesis beyond the pachytene spermatocyte stage. Recently, generation of Sertoli cell-specific knockout of the androgen receptor have clarified the roles of testosterone signalling by demarcating androgen signalling in these cells as essential for spermatogenesis but not for non-gonadal functions of testosterone. Interestingly, signalling through the androgen receptor has also been found to have important roles in female fertility as null females show longer estrous cycles and reduced fertility.

Studies from our group and several others have also discovered the essential roles played by multiple TGFß family members in reproduction. Among these, the *Inha* mouse model revealed unexpectedly important roles for the protein in gonadal cell-cycle regulation as null animals developed gonadal tumours of granulosa cell/Sertoli cell origin. As identified by our group and others, also important in reproduction are the activin signalling components; activin receptor type II mutants develop testicular hypoplasia and activin ß

B subunit mutants have defective parturition. In order to analyze the functional interaction between activin ß A and activin ß B, our group also created a knockin allele of ß B into the ß A locus. The knockin rescued the neonatal lethality of the ß A-knockout mice, thus allowing dissection of the reproductive phenotype (Brown *et al.* 2000). While male knockin mice had delayed fertility, female mice had severely reduced fertility, thus highlighting the important roles played by activin ß A in mammalian reproduction. Also significant is the follistatin conditional knockout model in which the activin/BMP/GDF antagonist follistatin is deleted only in the granulosa cells of the ovary. Female mice with a granulosa cell-specific deletion of follistatin have severely reduced fertility due to reduced numbers of ovarian follicles and defects in ovulation and fertilization. These and other models have identified an enormous number of proteins belonging to the TGFß super family with essential roles in reproduction.

Early Embryo: The Maternal Effect

In the early embryonic period prior to zygotic genome activation, maternally supplied proteins play critical roles in pre-implantation development. The first such maternally derived factor identified was Mater (*Nalp5*). Embryos lacking maternal NALP5 protein arrested at the two-cell stage and degenerated, thereby suggesting an essential role of maternal NALP5 protein in embryonic development. Maternally supplied oocyte-specific isoform of the DNA methyltransferase 1 (*Dnmt1o*) was also found to be essential for maintenance of methylation at imprinted loci in embryos; heterozygous embryos of null females died *in utero* due to alteration of allele-specific expression from imprinted loci. Methyltransferases that are implicated in initiation of methylation at imprinted loci (Dnmt3 1 and Dnmt3a) have also been demonstrated to be essential maternally supplied proteins for embryonic viability. Recently our group and others have added to the list of maternal effect genes in mice.

First, female mice null for an oocyte-specific gene *Zar1* (zygote arrest 1) were found to be infertile due to a failure of progression of fertilized embryos from the one- to two-cell stage; in the arrested embryos the parental genomes remain separated and there is a marked reduction in the synthesis of the transcription-requiring complex. Thus *Zar1* seems to act at an earlier time-point than *Mater* in embryonic development. Next, a mouse homolog of *Xenopus* nucleoplasmin (nucleoplasmin 2, *Npm2*) was identified to be critical in the process as well. Embryos from *Npm2*-null females had defects in nuclei and nucleolar organization specifically evident in the loss of heterochromatin from these organelles. Although, some embryos are able to proceed through the two-cell stage and diminished numbers of offspring are born from null females, most embryos show reduced cleavage to the two-cell stage. Another recent maternal-effect factor to be identified is Stella or *Dppa3* (developmental pluripotency-associated 3). Embryos from *Dppa3* $^{-/-}$ females progress to the four-cell stage but are defective in further cleavage and in pre-implantation development. Future experiments utilizing *in silico* approaches to identifying early-embryo-enriched genes would provide more candidate genes that might play important roles during this phase of embryonic development.

Conditional Mutagenesis

Although a knockout mouse is a valuable resource to analyze the effects of a null mutation in a tissue of interest, in a large number of cases the study of the phenotype is precluded by a more widespread effect caused by the mutation. Also, distinguishing between the cell-autonomous and cell-non-autonomous nature of action of a gene of interest is not possible in a ubiquitous knockout. Researchers have long recognized this drawback of the traditional knockout approach and have sought to develop technologies that would allow ablation of a gene function in a specific tissue and at a specific time: this is the rationale behind the conditional mutagenesis approach. Over the years refinement and innovation have led to a whole set of new tools with which conditional knockouts are achieved *in vivo* at present; here we discuss the advantages and pitfalls of the use of the binary Cre-*loxP* technology in the generation of conditional mutants within the context of reproduction.

The binary Cre-*loxP* method is so designed that the bacteriophage recombinase Cre is driven by a tissue- or stage-specific promoter in a restricted spatio-temporal manner; the deletion of gene function in that tissue is then dependent on the Cre recombinase to bind to its cognate recognition sequences (*loxP*) sites and effect a DNA recombination event between any two of such *loxP* sites, which can lead to deletion, inversion and other genomic rearrangements based on the position and orientation of such sites. The gene (or more commonly, the gene region) to be deleted is flanked by *loxP* ('floxed') sites in a targeting event similar to a conventional knockout; floxed mice are then bred with mice expressing Cre under the regulation of a promoter of choice to obtain tissue-specific knockouts in the offspring. Using this method several conditional knockouts have been generated that display reproductive defects; indeed, some of them were created after conventional knockouts resulted in embryonic or post-natal lethality (*Oct4* and *Fst*).

One apparent deficit of this technology in reproduction is the paucity of available Cre- driver lines that are specifically expressed in reproductive tissues of interest. Another problem relates to the fact that our restricted knowledge of gene-regulatory elements results in use of promoter regions that prove to be 'leaky', meaning that they drive Cre expression and consequently gene deletion outside of their specific spatio-temporal zone (for example the *Sf1*-Cre line), thus defeating the very purpose of creating a conditional knockout. The last problem, and not a trivial one, is the issue of the efficiency of Cre-mediated recombination. Cre lines often express the recombinase at levels that are not sufficient to achieve complete recombination between *loxP* sites, thus resulting in the generation of a hypomorph.

Attention has been focused on producing more efficient deleter Cre lines but we are still some distance from achieving mastery over this elegant but at times difficult technology. Nevertheless, with the intensive efforts of various groups and the advent of newer technologies, for example inducible Cre lines, the *Flp-Frt* system, it is conceivable that in the future, instead of one cure for all ills, the field will witness the use of a multitude of technologies each best tailored in a particular situation to deliver the proper result: that of generation of a null mutation in the tissue of interest.

In the Future

Worldwide, 15% of couples are unable to conceive within 1 year of starting unprotected intercourse. Although fertility defects may be the secondary manifestation of systemic illnesses (endocrinopathies, neurological disorders, anatomic malformations), which may or may not be treatable, often the defect is primary to the reproductive system and requires specialized evaluation of urologists and/or gynecologists and fertility specialists for possible treatment. However, even with advanced clinical tools for diagnosis of reproductive defects, about a quarter of clinical infertility cases are diagnosed as idiopathic. Since the identification of the genetic bases of a complex disorder such as infertility cannot be done through traditional genetic approaches of pedigree analysis and linkage studies, our understanding of the genetics of reproduction in humans has remained minimal and is limited to a few documented examples such as the discovery of Y-chromosome microdeletions in azoospermic men or the findings of cystic fibrosis transmembrane conductance regulator (CFTR) mutations in patients with congenital bilateral absence of the vas deferens. This raises the possibility that some of these 'idiopathic' cases are likely due to as-yet unidentified genetic anomalies.

In this review we have illustrated some examples of how knockouts in mice have greatly enhanced our understanding of this complex biological process. This is not meant to be a comprehensive survey of all knockouts with reproductive defects but rather an illustrative essay of the utility of the knockout approach towards understanding a process. Challenges remain plenty in both our understanding of the process and in the use of this technology itself, but the emergence of exciting technologies and projects such as RNA interference in mammals, the MICER project for genome-wide large-scale insertional mutagenesis from indexed targeting vectors, and the recent conceptualization of a whole-genome knockout mouse project, or KOMP, by an international collaboration will broaden the scope and increase the enthusiasm of researchers in the power of knockouts and knockdowns in evaluating the roles of additional proteins in the reproductive axis *in vivo*.

Viruses and Mammalian Reproduction

Remnants of ancient viral infections are found in the genomes of many species, including all mammals and humans. The genetic material that a virus leaves behind can disrupt genes causing cancer or diseases, but usually is harmless and sometimes it proves beneficial, providing protection from infection or being involved in reproduction. In particular, a class of retroviruses, known as endogenous retroviruses (enJRVSs) related to Jaagsiekte sheep retrovirus (JSRVs), are critical during the early phase of pregnancy when the placenta begins to develop, as a team of US and Scottish scientists has discovered on sheep.

Retroviruses are best known for their ability to cause diseases (e. g. HIV or JSRVs which provoke lung cancer on sheep). They are unique for their ability to permanently insert their genetic DNA into the DNA of host cells, said Thomas Spencer, a reproductive biologist with the Texas Agricultural Experiment Station and Texas A&M University. During mammalian evolution, some retroviruses infected the reproductive cells of the host, which is then inherited by their offspring. Many scientists believed these endogenous retroviruses were junk DNA, Spencer said. The idea that endogenous retroviruses are important for reproduction in mammals has been around for about

30 years, Spencer said. Studies in cultured cells have shown that a protein of a human endogenous retrovirus might have a role in the development of the human placenta.

The team blocked expression of the envelope protein of the enJSRVs using morpholino antisense oligonucleotides, which inhibit translation of specific messenger RNA, into sheep uteruses 8 days after fertilization—6 to 8 days before the placenta begins to form. By the 16th day after fertilization, the treated embryos were already smaller and more fragile than controls that were treated with random DNA fragments. None of the five treated embryos' placentas formed properly, and all but one died by day 20. The growth of the placenta was reduced and a certain cell type, termed giant binucleate cells, did not develop.

The result was that embryos could not implant and the sheep miscarried, Spencer said. Miscarriage is a serious medical problem for all mammals, including humans.

"Our research supports the idea that endogenous retroviruses shaped the evolution of the placenta in mammals and then became indispensable for pregnancy, and thus may be why they are expressed in the placenta of many mammals," he said.

"The enJSRVs arose from ancient infections of small ruminants during their evolution," said. Dr. Massimo Palmarini, a virologist at The University of Glasgow Veterinary School. "This infection was beneficial to the host and was then positively selected for during evolution. In other words, animals with enJSRVs were better equipped than those without. Therefore, enJSRVs became a permanent part of the sheep genome and, in these days, sheep can't do without them."

The placenta forms when cells in the outermost layer of the embryo fuse with those lining the walls of the uterus. In many mammals, the placenta is loaded with certain viral envelope proteins, leading scientists to suspect that these proteins may be essential for its development. Indeed, experiments done with human and mouse cells in culture show that these proteins can make cells fuse with each other, suggesting that they may be responsible for the fusion of cells in the developing placenta.

"This is the first study done within animals that strongly supports the idea that endogenous retroviruses have become essential for placental development and thus successful reproduction of some mammals" says Spencer, whose team reports its work online this week in the Proceedings of the National Academy of Sciences (PNAS).

"Because similar viruses are present in the placentas of a wide range of mammals, including humans, the placenta might have evolved multiple times during evolution with different viruses playing a role", says study author and veterinary pathologist Massimo Palmarini of the University of Glasgow Veterinary School in the U.K.

Until this study, there were only "correlations and suspicion" that viral gene products could be involved in the development of the placenta, but nothing had been proven, says R. Michael Roberts, a developmental biologist at the University of Missouri, Columbia. Spencer hopes that future studies will reveal exactly how the envelope protein regulates the growth and development of the placenta, with implications for both human health and animal production.

Mammalian Reproduction: An Ecological Perspective

The objectives of this paper are to organize our concepts about the environmental regulation of reproduction in mammals and to delineate important gaps in our knowledge of this subject. The environmental factors of major importance for mammalian reproduction are food availability, ambient temperature, rainfall, the day/night cycle and a variety of social cues. The synthesis offered here uses as its core the bioenergetic control of reproduction. Thus, for example, annual patterns of breeding are viewed as reflecting primarily the caloric costs of the female's reproductive effort as they relate to the energetic costs and gains associated with her foraging effort. Body size of the female is an important consideration since it is correlated with both potential fat reserves and life span.

Variation in nutrient availability may or may not be an important consideration. The evolutionary forces that have shaped the breeding success of males usually are fundamentally different from those acting on females and, by implication, the environmental controls governing reproduction probably also often differ either qualitatively or quantitatively in the two sexes. Mammals often live in habitats where energetic and nutrient challenges vary seasonally, even in the tropics. When seasonal breeding is required, a mammal may use a predictor such as photoperiod or a secondary plant compound to prepare metabolically for reproduction.

A reasonable argument can be made, however, that opportunistic breeding, unenforced by a predictor, may be the most prevalent strategy extant among today's mammals. Social cues can have potent modulating actions. They can act either via discrete neural and endocrine pathways to alter specific processes such as ovulation, or they can induce nonspecific emotional states that secondarily affect reproduction. Many major gaps remain in our knowledge about the environmental regulation of mammalian reproduction.

For one, we have a paucity of information about the annual patterns of breeding and about the mechanisms controlling these patterns in the most common mammals on the planet-the small to average-sized mammals living in the tropics. We probably have only a shallow conceptualization of the way available energy and nutrients control reproduction and, likewise, we may have only a narrow view of the potential kinds and uses of seasonal predictors. Finally, we have little appreciation of the way environmental cues interact with each other to control reproduction.

Rut (Mammalian Reproduction)

The rut is the mating season of ruminant animals such as deer, sheep, elk, moose, caribou, ibex, goats, pronghorn and Asian and African antelope.

During the rut (also known as the rutting period, and in sheep sometimes as *tupping*), males often rub their antlers or horns on trees or shrubs, fight with each other, wallow in mud or dust, and herd estrus females together.

The rut in many species is triggered by a shortening of the length of daylight hours each day. The timing of the rut for different species depends on the length of their gestation period (length of pregnancy), usually occurring so the young are born in the spring, shortly after new green growth has appeared (which provides food for the females, allowing them to provide milk for the young), and when the temperatures are warm enough that the young will not die of hypothermia.

White-tailed Deer

The rut for white-tailed deer (*Odocoileus virginianus*) usually lasts from 1-3 months in the Northern Hemisphere and may occur most of the year in tropical zones. The rut is the time when white-tailed deer, especially bucks, are more active and less cautious than usual. This makes them easier to hunt, as well as more susceptible to being hit by motor vehicles.

Some people believe that the white-tailed deer rut is also controlled by the lunar phase and that the rut peaks seven days after the second full moon (the rutting moon) after the autumnal equinox on September 21. However, study of white-tailed doe conception dates conducted in

Minnesota between 1980 and 1987 showed no correlation between peak breeding dates of white-tailed deer and any lunar phase.

A white-tail doe may be in estrus for up to 72 hours, and may come into estrus up to seven times if she does not mate. Cow elk may come into estrus up to four or more times if they do not mate.

Embryonic Development: Getting Started

In animals, one can usually distinguish 4 stages of embryonic development:

- Cleavage
- Patterning
- Differentiation
- Growth.

Cleavage

Mitosis and cytokinesis of the zygote, an unusually large cell, produces an increasing number of smaller cells, each with an exact copy of the genome present in the zygote. However, the genes of the zygote are not expressed at first.

The early activities of cleavage are controlled by the mother's genome; that is, by mRNAs and proteins she deposited in the unfertilized egg.

In humans, the switch-over occurs after 4—8 cells have been produced; in frogs not until thousands of cells have been produced. Cleavage ends with the formation of a blastula.

Patterning

During this phase, the cells produced by cleavage organize themselves in layers and masses, a process called gastrulation. The pattern of the future animal appears:

- front and rear (the anterior-posterior axis)
- back side and belly side (its dorsal-ventral axis)
- left and right sides.

There is little visible differentiation of the cells in the various layers, but probes for cell-specific proteins reveal that different groups of cells have already started on specific paths of future development. Gastrulation forms three major "germ layers": ectoderm, mesoderm, and endoderm. By gastrulation, the genes of the zygote genome are being expressed.

Differentiation

In time, the cells of the embryo differentiate to form the specialized structures and functions that they will have in the adult. They form neurons, blood cells, skin cells, muscle cells, etc., etc. These are organized into tissues, the tissues into organs, the organs into systems.

Growth

After all the systems are formed, most animals go through a period of growth. Growth occurs by the formation of new cells and more extracellular matrix.

The Problem

The genome of the zygote contains all the genes needed to make the hundreds of different types of cells that will make up the complete animal. There are two major categories of these genes:

- "housekeeping" genes = those that encode the RNAs and proteins needed by all kinds of cells. Examples:
  - genes for tRNAs, rRNAs
  - enzymes of glycolysis.
- tissue-specific genes = those that encode mRNAs and hence proteins that are used by one or a few specific kinds of cell. Examples:
  - genes for hemoglobin expressed in the precursors of red blood cells
  - the gene for insulin expressed in the beta cells of the islets of Langerhans.

However, every cell descended from the zygote has been produced by mitosis and thus contains the complete genome of the organism (with a very few exceptions).

Two pieces of evidence:

- Dolly. Dolly is the sheep that was formed by inserting a nucleus from a single cell of an adult sheep into an enucleated sheep egg. She proves that the cell from the adult had lost none of the genes needed to build all the tissues of a sheep.
- Spemann's egg-tying experiments. Many years earlier, the German embryologist Hans Spemann demonstrated the same truth. He used strands of baby hair to tie loops around fertilized salamander eggs. Although the egg half with the nucleus began cleaving normally, the other side did not begin cleavage until

a nucleus finally slipped through the knot. So long as the egg was tied so that both halves contained some of the gray crescent, the second half began normal cleavage and ultimately produced a second tadpole. Even after 5 mitotic division of the zygote nucleus (the 32-cell stage), the entire genome was still available in each descendant nucleus.

SO, How do the various differentiated cells come to express the appropriate subset of tissue-specific genes and not all the genes of the genome?

The Solution

1. A fertilized egg is much larger than the normal cells of an animal's body. Some (e.g., a hen's egg) are truly huge. The frog egg has a volume 1.6 millions times larger than a typical frog cell. The photo (courtesy of L. M. Beidler) is of a 16-cell frog embryo. This mass of cells is no larger than the original egg. The eggs of mammals are smaller, but even they are larger than their descendant cells will be.
2. The cytoplasm of the fertilized egg is not homogeneous. It contains gradients of mRNAs and proteins. These are the products of the mother's genes and were deposited in the egg by her.
3. Cleavage of the fertilized egg partitions it into thousands of cells of normal size. Each contains a nucleus descended from the zygote nucleus.
4. But each nucleus finds itself partitioned off in cytoplasm containing a particular mix of mRNAs and proteins.
5. When the frog blastula has produced some 4,000 cells, transcription and translation of its nuclear genes begins (and the mother's mRNA molecules, that up to now have been the source of all protein synthesis, are destroyed).
6. The genes that are expressed by the nucleus in a given cell are regulated by the molecules, mostly protein transcription factors, found in the cytoplasm surrounding that nucleus.
7. Once a cell-specific pattern of gene expression is launched, that cell may release molecules that regulate the genes of nearby cells.
8. In this way, the foundation is laid for the building of an organism with hundreds of types of differentiated cells — each in its correct location and performing its correct functions.

First Example: Xenopus

- During egg formation, molecules of mRNA encoding the protein VegT are deposited at the vegetal pole of the cell.
- Cells that form there during cleavage translate the mRNA into the VegT protein.
- VegT is a transcription factor that turns on genes that produce members of the transforming growth factor-beta (TGF-â) family (e.g., activin).
- These proteins are needed for cells to start down the path to becoming mesoderm.
- Some of those cells will, in turn, become the Spemann organizer.
- Later, the Spemann organizer will secrete molecules that induce the ectodermal cells above them to develop into the tissues of the brain and spinal cord.

A Demonstration

The photo on the left (courtesy of Douglas Melton, in whose lab the work was done) shows the tadpole that developed from a frog embryo that had been injected with mRNA for activin.

At the 32-cell stage of cleavage, a single cell— that would normally have gone on to form part of the belly of the animal — was injected with activin mRNA. A two-headed tadpole resulted (arrow). (A normal tadpole is shown on the right for comparison.) The mother is also responsible for making ectoderm.

During egg formation, she deposits mRNAs encoding Ectodermin at the animal pole of the egg. Ectodermin interferes with SMAD signaling and thus prevents TGF-â proteins from forming mesoderm at the animal pole.

Second Example: Drosophila

In Drosophila and other insects, cleavage involves repeated mitosis but without cytokinesis. So the daughter nuclei remain suspended within the single egg compartment.

After several thousand nuclei have been formed, they migrate to the margins of the egg. Only then do plasma membranes form around each nucleus forming true cells. But, as for Xenopus, the genes that will be expressed by those cells are regulated by the cytoplasmic constituents they found themselves surrounded by. And that, again like Xenopus, is determined by where those molecules end up in the egg.

Bicoid

For example, Drosophila eggs have a gradient of mRNA transcribed from a gene designated *bicoid (bcd)*. The transcripts are deposited in the egg by "nurse" cells surrounding it.

Once within the egg, they are transported (along microtubules) toward the anterior. The result is a concentration gradient of bicoid mRNA extending from a high level at the anterior of the egg to a low level at the posterior. After fertilization, the mRNAs are translated into bicoid protein. High levels of the protein lead to the formation of the head of the larva.

Nanos

Conversely, the posterior of the egg has a high concentration of mRNA encoding the nanos protein, which is needed to form the structures of the tail of the larva.

Three demonstrations:

1. Remove some of the bicoid-rich cytoplasm from the anterior of the fertilized egg and replace it with nanos-rich cytoplasm from another egg. The result: a larva (nonviable) with a tail at each end.
2. Inject the anterior of the fertilized egg with nanos mRNA. The result: another double-posterior larva.
3. Make female fruit flies that are transgenic for a recombinant gene containing:
   - the structural gene for nanos
   - coupled to the 3′ anterior-directing signal of the bicoid gene.

Third Example: The Mud Snail

The mud snail, Ilyanassa obsoleta, is a small gastropod that lives in mud flats along the Atlantic coast.

Like other protostomes, cleavage of the zygote produces daughter cells that are already committed to their fate. In other words, even as early as the two-cell stage, the cells are no longer totipotent. Unlike humans and other deuterostomes, then, identical twins cannot form.

In the 12 December 2002 issue of Nature, J. David Lambert and Lisa Nagy reported another mechanism by which two daughter cells become committed to different fates even though they have inherited the same genome.

They traced the distribution in the cells of early embryos of the messenger RNAs (mRNAs) encoding 3 proteins that are known to be important in the development of other animals such as Xenopus and Drosophila.

- *IoEve*, which is Ilyanassa obsoleta's version of *even-skipped* (*eve*) in Drosophila;
- *IoDpp*, which is the snail's version of
  - o *decapentaplegic* (*dpp*) in Drosophila and the genes encoding
  - o bone morphogenic proteins (BMP2 and BMP4) in vertebrates
- *IoTld*, which encodes the snail's version of a protein called tolloid in Drosophila.

One of the first events when an animal cell prepares to divide by mitosis is the duplication of its centrosome and separation of the duplicates.

Lambert and Nagy found that :

- in interphase the messenger RNAs were distributed diffusely throughout the cytosol, but
- as the cell got ready for cleavage, the mRNAs collected at only one of the now pair of centrosomes. They were collected there by traveling along the microtubules that radiate out from the centrosome.
- As cleavage continued, the mRNAs moved from the centrosome to a spot on the inner surface of the plasma membrane. They got there by travelling along actin filaments.
- At cytokinesis, this patch of accumulated mRNAs was incorporated exclusively into the smaller daughter cell.

Centrosome sorting (of proteins in this case) also plays a role in determining whether embryonic cells of Caenorhabditis elegans remain in the germline or become the somatic cells of the worm.

Embryonic and Genetic Manipulation in Fish

Lives on earth first appeared over 3 billion years ago. As a result of evolution, vertebrates are the utmost advanced life form distributing from seabed to high-mountain. Fishes, at the lower stage of evolution but being the biggest community in vertebrates, include about 21,700 to 28,000 species that take over almost half of the total number of vertebrates. Fishes are also models for experimental study as the eggs are large, fertilization and development are externally and the embryos are transparent. In addition, mono-sexual breeding and crossbreeding between many distantly related species could be done in fishes.

Fish culture is one of the earliest activities of human civilization. Time dating back to 2500 years ago when China was at the Spring and Autumn Period, an ancient Chinese named Fan Li created a great literature "Handbook of Fish Culture" in which domestication and cultivation of common carp in ponds were described in detail.

This is the first monograph of fish culture in the world. Since then, Chinese farmers enjoyed a long tradition of selective domestication resulting in very valuable "four farming species" of fish for pond culture. These are black carp (Mylopharyngodon poceus), grass carp (Ctnopharyngodon idellus), silver carp (Hypophthalmichihys molivrix) and big-head carp (Aristichthys nobilis). Based on the feeding habits of these fishes, Chinese farmers also invented "multi-culture method" of stocking the four farming species together in the same water body. This is in fact a sustainable ecological system by that farmers are able to gain maximum output with minimum cost. These domesticated species and the multi-culture method have been favourably adopted by other nations abroad.

In addition to the traditional experiences of fish farming, scientists in China initiated some fundamental research in fish cell and developmental biology and molecular genetics leading to a breakthrough in fish biotechnology. The major achievements are pointed as follows. In 1960's, Professor Tung T. C. first introduced the art of nuclear transplantation with fish to study the interaction between nucleus and cytoplasm. Tung and his colleagues succeeded in generation of "nuclear-cytoplasmic hybridized fish" between different species (reviewed in). In early 1980's, the first somatic cell cloned fish was derived from kidney cells by nuclear transplantation. In 1984, the first batch of transgenic fish was also generated. The development and combination of technologies of nuclear transplantation and gene transfer in fish predicate that a new era in fish breeding is coming; fish directional breeding will revolutionize the traditional fish farming industry.

Nuclear Transplantation in Fish

The concept of nuclear transplantation dates back to 1938 when a German biologist, Hans Spermann, proposed an experiment that would evaluate the relative importance of the nucleus and cytoplasm in controlling early developmental events. In the early 1950's, the technology of nuclear transplantation was first demonstrated with frogs by Briggs and King and amphibians were becoming most commonly used for studying nuclear transplantation (reviewed in).

These studies revealed that the nuclei undergo restriction in developmental capacity as cells became differentiated. Although several experiments suggested that differentiated somatic cells still had developmental totipotency, the cell differentiation didn't necessitate any irreversible changes in nuclear genetic material. Some subsequent experiments with differentiated somatic cells from Xenopus, and Ranna pipens could hardly supported the previous cases because that nuclei from these cells only supported the ontogenesis to feeding-stage of tadpole but not to an adulthood. Thus, whether ultimately differentiated cells have the capacity to support the reconstructed embryos to complete their full ontogenesis still remains unanswered.

Illmensee and Hoppe in 1981 first succeeded in nuclear transplantation in mammals. They reported that an inner cell mass nucleus could support early development of an enucleated egg in mouse. Three years latter, McGrath and Solter reported nuclear transplantation in mice by using micrioinjection coupled with cell fusion technique. Since then, embryonic nuclear transplantation in sheep, calve, rabbit, porcine and goat were successively done. In 1997, the birth of cloned lamb "Dolly" derived from an adult mammary gland cell marked the beginning of a "golden-age" in animal cloning. Recently, success has also been announced in cloning mice, goat, and cattle derived from differentiated cells: fetal fibroblasts, muscle cells, cumulus cells, or oviductal cells. Successes on somatic cloning conclusively revealed that differentiated adult cells still remain totipotent and maintain the whole genome to support normal development to term.

In amphibian and mammals, nuclear transplantation has been successfully done within the same species. In other words, the donor of nucleus and the host of enucleated eggs must come from the same species. On the other hand, "Tung's fish" brought an absolutely new story. Tung's embryonic nuclear transplantation in fish involves three steps in general,. (1) Preparation of donor cells: the blastoderm was separated from the yolk with a fine glass needle, and carefully placed in Holtfreter's dissociation solution for further separating into individual cells. (2) Preparation of unfertilized host eggs: the mature eggs stripped from the female were dichorionized with a pair of forceps, and enucleated by inserting a sharp glass needle into the egg cytoplasm just underneath the site of the second polar body. (3) Nuclear transplantation: after donor cell was sucked into the micropipette by a slight negative pressure and microinjected into the enucleated eggs, the reconstructed embryos were put into Holtfreter's solution for further development. Great successes of fish nuclear transplantation were achieved between different genus (common carp (Cyprinus Carpio Linnaeus) and crucian carp (Carassius aurantus), and different subfamilies (grass carp (Ctenopharyngodon idellus) and blunt-snout bream (Megalobrama amblycephals)),. The nuclear-cytoplasmic hybrid fish revealed that while most phenotypic characteristics were controlled by the nucleus, a few were controlled by the cytoplasm or by both (reviewed in).

In 1984, the first somatic cell cloned fish derived from short-term cultured kidney cells of triploid crucian carp was produced. In this experiment two rounds of nuclear transfer were carried out. In the first round, the nucleus (3n) was transferred into the enucleated eggs (2n) of crucian carp. There were 41% of the injected eggs developed into blastulae but no future development occurred. Nuclei from the blastulae were taken for second round of transfer into another batch of enucleated eggs (2n). In the second round, 8 gastrula generated and one of them developed into a fertile female fish (1.2%) with normal morphological features of crucian carp. The chromosome number was triploid (3n=150). This was the first case of cloned animal with somatic cells in fishes. It suggested that some nuclei of somatic cells, following two rounds of nuclear transplantation, could be reprogrammed to totipotent status as zygotic nucleus does.

Nuclear transplantation in fishes, as an approach for studying the relative roles of nucleus and cytoplasm in controlling the characteristics, had also been considered as a helpful method in obtaining new farming strains of fishes. Yan indicated that the nuclear-cytoplasmic hybrid was 22% higher in growth rate, 3.8% higher in protein content and 5.58% lower in lipid content than that of the control, respectively. The art of nuclear transplantation in fish, however, not only requires very skilful manipulators but also bring about very low efficiency in producing the nuclear-cytoplasmic hybrids. The application of nuclear transfer fish in fishery is, thus, under restriction to a certain extent.

Gene Transfer in Fish

People never give up their effort on pursuing of breeding new strains of farming species with high quality. Meanwhile, researchers are seeking for some genetic materials other than the whole cell nucleus for transfer. Around the late 1970's, researchers in China introduced the genomic DNA of common carp into the fertilized eggs of Mud carp, a tropical species. About 8% of the "total DNA-transferred" founder Mud carp showed increase of cold-resistance.

In early 1980's, with the advancement of techniques in molecular cloning and embryonic micro-manipulation, recombinant genes are able to be constructed and transferred into the host animals. The transgenic "super mouse" was the most stimulating report in transgenic studies.

The first batch of faster growing transgenic fish was generated by introducing a recombinant human growth hormone (hGH) gene capped with a mouse metallothionein-1 (MT) gene promoter into goldfish in 1984. In this study, fertilized eggs of goldfish were obtained by artificial spawning and insemination and the chorion of eggs were removed by digestion in 0.25% trypsin solution. About 1-2nl of DNA solution containing about 10^5–10^6 copies of the MThGH gene was delivered with a micromanipulator into the germinal disc just underneath the second polar body. All the manipulated eggs were carefully put into Holtfreter's solution for further development.

By Southern hybridization, more then 50% of the founders are transgenics. Since then, dozens of laboratories all over the world began to show great zeal for the study of transgenic fish, and gene transfer into fish embryos were performed in several species, such as rainbow trout. Atlantic salmon, tilapia (Oreochromis nilotica), medaka, common carp, zebrafish, loach (Misgurnus anguillicaudatus), catfish (Parasilurus asoltus Linnaeus), etc. Some other techniques, e.g. electroporation and sperm-mediating,, were also successfully employed in producing of transgenic fish. In addition to gene transfer with GH gene, other types of genes were also employed in gene transfer in fish. For example, GFP (green fluorescent protein) gene was transferred into fertilized eggs of zebrafish as a reporter and AFP (antifreeze proteins) gene was introduced into Atlantic salmon to gain freeze-resistant salmon. However, gh is so far the most commonly used and most thoroughly investigated type of transgenes. It was in a transgenic fish model that the behaviour of a foreign gene, this time the MThGH gene, in embryogenesis was intensively studied by Zhu et al. Southern blotting revealed that the behaviours of MThGH gene in host fish were a dynamic process, including replication, degradation, concatemer formation, and integration during embryogenesis. The replication began at very early stage of cleavage and suddenly took place at late-blastula to early-neurula stages. After neurula stage, most of the foreign gene were migrating with the host chromosomal DNA as revealed on the agarose gel by electrophoresis. It was suggested that foreign gene at these stages was in form of either large concatemers situated outside the chromosomes or integrated into the host genome. If the integration occurred at germ line cells of the host fish, the transgene could be inherited to their offspring. Northern hybridization showed that the transcripts of hGH gene could only be found at post-late-gastrula stage, which was consistent with the timing of the differentiation of fish embryonic cells. As a result of the expression of MThGH gene, transgenic fish showed significantly faster growing trait.

On the other hand, a certain proportion of the founders did not show growth enhancement and a few even grew slower than the control. This unexpected observation was reasonable when the multi-sited integration and transgenic mosaicism were taken into account. In the case of the generation of transgenic mammals, foreign gene had been microinjected into the zygotic pronucleus, which resulted in the integration of foreign gene occurring mostly before the first cleavage.

It was believed that the status of transgene in founder mammals was in a homozygous form, which could be transmitted to next generation in a Mendelian manner. However, pronucleus of fish eggs is not visible and the foreign gene could only be microinjected into their cytoplasm. The foreign gene integration in fish was found to be spanned over a long time course from gastrula to late developmental stages resulting in multi-sited integration and consequent transgenic mosaicism. Three categories of transgene integration could be deduced: functional integration, silent integration, and toxic integration. It is only in the functional integration that the transgene was integrated into host genomic sites suitable for expression and therefore showing "fast-growth" trait. Thus, the founder generation (P_0) of transgenic fish was far from a genetically homogeneous strain.

The specific growth rate (SGR) of P_0, F_1, and F_2 MThGH-transgenic fish, as well as of F4 generation was significantly higher than that of the controls,. Bioenergetic analysis on MThGH-transgenic fish compared with controls was thoroughly worked out. When feeding with fresh tubificid worms, the energy budget of both transgenic and control fishes can be expressed by the following equations:

MThGH-transgenic

$$F_2 \text{ fish } 100C = 8.9\text{ F} + 0.63\text{U} + 49.03\text{ R} + 41.44\text{ G}$$

$$\text{Control fish } 100C = 7.37\text{ F} + 1.14\text{ U} + 53.36\text{R} + 38.13\text{G}$$

In which, C is the total energy from food, F is the energy lost in faeces, U is the energy lost in nitrogenous exaction, R is the energy channelled to metabolism and G is the energy channelled to growth.

Compared with the controls, transgenic fish had a significantly higher proportion of food energy channelled to G and a significantly lower proportion of that channelled to R and U. The transgenic fish saved 6.62% of the total energy from eaten food for growth improvement. That phenomenon was named as "fast-growing and less-eating" effect.

Growth and feed utilization by MThGH-transgenic F4 fish feeding with diets containing different protein levels had also been carried out. Protein and energy intakes were significantly higher in the transgenics than in the controls fed with 20% protein diet, and recovered energy, as a proportion of protein intake, was also significantly higher in the transgenics than in the controls fed with 40% protein diet.

It was thus concluded that at a lower dietary protein level, transgenics achieved higher growth rates mainly by increasing food intake; but at a higher dietary protein level, transgenics achieved higher growth rates mainly through higher energy conversion efficiency. That is to say, transgenics are more efficient in utilizing dietary protein than the controls, which leads to transgenics getting a significantly higher specific growth rate than the controls. For the body composition, it was revealed that the transgenic fish had body contents of dry matter 1.6%, and protein 2.2% to 4.3% higher than those of the controls, but contents of lipid 3.9% to 13.1% lower than that of the controls.

The apparent digestibility of amino acids tends to be higher in the transgenics than in the controls, especially in fish fed diets with lower protein levels. While taking a look at the proportion of amino acids in transgenics and controls, there was no difference whatever betwen 17 amino acids. Thus, transgenic fishes would have much more nutritious value than control fishes. It is reasonable that fishes with "fast growing and less eating" as well as "high protein content and low lipid content" traits will fulfil human increasing requirement on protein source from fishes.

Nevertheless, both the mouse metallothionein-1 (MT-1) gene promoter and the hGH structural gene are not suitable for the purpose of producing farming species of transgenic fish. It was urged to construct "all-fish" gene for transfer. Researchers have cloned both common carp actin gene (CA), and grass carp growth hormone gene (gcGH), and subsenquetly made a new construct of pCAgcGH, an "all-fish" genomic construct with a powerful promoter of b-actin gene from common carp and the whole transcription unit of GH gene from grass carp. In the spring of 1997, this construct has been microinjected into the fertilized eggs of Yellow River Carp (Cyprinus Carpio L.), and a batch of CAgcGH-transgenic was produced. As these fishes grew up to 5-month-old, body weight of the heaviest transgenic individual was 2.75 kg, while that for the controls was 1.1 kg. About 10% of the transgenics were over 2.0 kg, while the controls were about 0.7 kg on average. It is more exciting that the heaviest body weight of 17-month-old transgenics, 7.65 kg, was two fold and more than that of the control siblings [unpublished data]. Some experiments revealed further that the gene constructs with genetic elements derived from fish expressed more efficiently in fish cells than that from mammalian sp., e.g. the transcriptional activity of mouse MT promoter was only 1/2 of that of carp MT promoter in CAT (chloramphenicol acetyl-transferase) gene-transferred fish cells. Till now, more than ten "all-fish" recombinant genes have been constructed all over the world. The most dramatic growth acceleration came from Delvin et al., in which their pOnMTGH1-ransgenic salmon were more than 11-fold heavier than the controls on average.

It is very sensitive to talk about the biosafety of transgenic fish. There are food, genetic and ecological safeties, each of which should be concerned seriously. At present, the widely accepted principle on safety evaluation of foods produced by modern biotechnology was the "substantial equivalence principle" delivered by European OECD (Organization for Economic Cooperation and Development) in 1993. According to this principle, "all-fish" transgenic carp is included in "level I", the safest level. Studies from Cui's investigation revealed that the transgene could only flow among individuals within a species but not between species by natural reproduction. Fish crossbreeding has been widely used in aquaculture, two sets of whole genome of different species mixing with each other, i.e. 10^5 genes of one species crossing with 10^5 genes of another species. On the other hand, the "all-fish" gene transfer can also be regarded as a crossbreeding, but 10^5 genes from one species cross with 1 gene of another species instead.

It can be simply figured out that the hetrozygosity of genome in the CAgcGH-transgenic common carp is about 105 times less than that in the hybrids between common carp and grass carp. In other words, the risk of stocking "all-fish" gene transgenics is considerably on a lower level in comparison with stocking the hybrid fish. Nevertheless, people should take a cautious attitude towards the application of transgenics, since transgenic fish is still a "newborn" in comparison with naturally existing species that have undergone a long evolutionary and selective course. As polyploid-breeding in fish has been very popular during the past, this technique can also be employed in the breeding of transgenic fish. By crossing tetraploid individual with haploid transgenics, the infertile triploid strain of transgenic fish could be generated. It is reasonable to consider that stocking infertile strain of transgenic fish will lessen their impact on water ecosystem to the least degree.

The Prospect of Transgenic Fish

Just as what has been discussed at the preceding part of this article, the transgenic fish generated at present is far from a genetic homogenous strain. Gene transfer in fish has not succeeded in the site-specific integration, controllable expression and stable transmission of the transgene. One of the most efficient ways to solve this problem is to use gene targeting technique in fish gene transfer. Gene targeting, homologous recombination between DNA sequence residing in the chromosome and newly introduced cloned DNA sequence, allows the transfer of any modified gene into the host genome of living cells. Since embryonic cell culture and embryonic cell nuclear transplantation, techniques have been developed in fishes, gene targeting is hopeful to be carried out to gain embryonic cell lines carrying artificially modified and site-specific integrated gene. By nuclear transplantation with gene-targeted embryonic cells, the genetic homogenous strain of transgenic fish can be generated. Additionally, to establish stem cell like lines in some model fish species, such as zebrafish and medakafish, were successfully reported. If researchers can make great progress on the study of stem cell in farming fishes, it will be more convenient to generate homogenous strain of transgenic fish for aquaculture.

4

Genetics Embryology

Embryology

Embryology is the branch of developmental biology that studies the beginning and early growth of sexually reproducing organisms. This life science speciality focuses primarily on *embryogenesis* (the formation and development of embryos). In humans, the term embryo is traditionally reserved for the first two months of development. After that point, the term "embryo" is replaced by the term "fetus," which then applies until birth.

At the beginning of the individual's development, the entity is a single cell. After two months, it has limbs, distinct fingers and toes, internal development, and countless cells. So the term "*embryo*" applies to an individual throughout a vast range of developmental change.

Since the time Charles Darwin published *The Origin of Species*, embryology has been used to support common descent. In fact, Darwin referred to embryological homology as the *strongest single class of facts* that existed to support his theories.

A theory later put forth by Ernst Haeckel, known as the biogenetic law, asserted that the evolutionary history of an organism was recapped during embryo development.

Although the biogenetic law is now discredited, in recent years, embryology has reemerged as a tool used by evolutionary biologists that attempt to establish phylogenetic relationships by identifying developmental similarities between taxonomic groups. *Evolutionary Developmental Biology* is a merging of developmental biology and evolutionary biology that is commonly known as "*evo-devo*".

Embryogenesis

Embryogenesis begins when a sperm fertilizes an egg and creates a single cell that has the potential to form an entire organism. In the first hours after fertilization, this cell divides into identical cells. Approximately 4 days after fertilization and after several cycles of cell division, these cells begin to specialize, forming a hollow sphere of cells, called a blastocyst. The blastocyst has an outer layer of cells, and inside this hollow sphere, there is a cluster of cells called the inner cell mass. The cells of the inner cell mass will go on to form virtually all of the tissues of the human body.

Although the cells of the inner cell mass can form virtually every type of cell found in the human body, they cannot form an organism. Therefore, these cells are referred to as pluripotent, that is, they can give rise to many types of cells but not a whole organism. Pluripotent stem cells undergo further specialization into stem cells that are committed to give rise to cells that have a particular function. Examples include blood stem cells that give rise to red blood cells, white blood cells, and platelets, and skin stem cells that give rise to the various types of skin cells. These more specialized stem cells are called multipotent—capable of giving rise to several kinds of cells, tissues, or structures.

Cleavage

After fertilization, the zygote proceeds immediately to the first cleavage (cell division) and subsequent cell divisions follow rapidly. The zygote is a very large cell, but the first waves of rapid cell division occur without increase in cell volume. The result is a closely bound mass of cells each of more typical cell size. At this stage the cells are called blastomeres, and the organism as a whole is called a morula from the time it has 16 blastomeres to the next stage. As it is undergoing this very rapid cell division, the organism is also migrating down the uterine tube toward the uterus.

Vertebrates, such as humans, have bilateral symmetry and therefore polarity in three dimensions (head-tail, or back-front, and left-right). Establishing polarity is one of the most basic manifestations of emerging specialization. But the egg is roughly spherical, and it is not readily apparent how polarity is established. Although it had been shown long ago that the point of sperm entry determines the plane of first cleavage (and thus subsequent ones) in amphibian eggs, mammals were believed until recently to remain spherically symmetrical until later in development. Recent data on mammalian zygotes, however, suggests that the point of sperm entry may similarly determine the cleavage plane. Even the first two cells resulting from the first cleavage may have different propensities, which persist through the next divisions as the progeny of one cell tend to become the body of the offspring and progeny of the other cell become the embryo's contribution to the placenta and other supporting structures. The word "fate," however, might be too strong, because the cells of such very early embryos are resilient to perturbations—if one cell is removed, the remaining ones can compensate.

Gastrulation

Gastrulation begins when fluid starts to accumulate between the blastomeres of the morula. The fluid-filled spaces run together, forming a relatively large fluid-filled cavity. At the point when the cavity becomes recognizable, the organism is called a blastocyst. The outer cells of the blastocyst, especially those around the blastocyst cavity, assume a flattened shape. The flattened cells of the exterior blastocyst are the trophoblast.

They become the embryo's contribution to the placenta and other supporting structures. On one side of the blastocyst is a group of cells that project inside into the blastocyst cavity; this is the inner cell mass, or embryoblast, and its progeny form the body of the new offspring.

The cells of the inner cell mass can give rise to progeny differentiating into all the types of cells in the adult body, so they are called pluripotent. They have not usually been described as totipotent because, the inner cell mass having already differentiated from trophoblast, the cells of the inner cell mass were believed to be no longer able to give rise to the cells of the trophoblast.

Recent work, however, describes culture conditions under which human embryonic stem cells can differentiate to trophoblast cells. Although the new offspring itself develops only from the inner call mass, the trophoblast is not just passive padding. Its progeny are the essential and specialized connection between the embryonic and maternal systems. Embryonic stem cells can be isolated from the inner cell mass.

Organogenesis

The basic structures and relations of all the major organ systems of the body emerge during the fourth through the eighth weeks of embryonic development. First, the embryo folds in several ways so that the flat linear structure distinguished by neural tube flanked by somites become roughly C-shaped. The effect of this is to bring the regions of the brain, gut, and other internal organs into their familiar anatomical relations. During the fourth week the neural pores, the ends of the neural tube "zipper," close. First the one at the cranial or head end, which is called the anterior or rostral pore, closes, and later the caudal or tail-ward pore closes. Closure of the neural pores completes the closure of what will become the central nervous system. Also during the fourth week, limb buds become visible, first buds for arms and later for legs. Further, two accumulations of cells along the neural tube become distinguishable: the alar plate and the basal plate. Cells of the alar plate go on to become mostly sensory neurons, while basal plate cells give rise mostly to motor neurons. Already while the neural tube is closing, its walls along the cranial area are thickening to form early brain structure. Cranial nerves, for example the nerves for the eye and for the muscles of the face and jaw, also are beginning to develop at this time. The embryonic brain develops rapidly in both size and structure especially during the fifth week, and the optic cup that will form the retina of the eye becomes visible as well.

Support for Darwinism

Von Baer's Law

Several decades before Darwin published the *The Origin of Species*, a German embryologist Karl Ernst von Baer had shown that the embryos of some vertebrates looked very similar during particular stages of development. This observation became known as *von Baer's law*. After reading a summary of von Baer's law in 1842, Darwin concluded that embryonic resemblances were a very strong argument in favour of common descent. In The Origin of Species (1859) Darwin stated *"Community of embryonic structure reveals community of descent"*, and in a 1860 letter to Asa Gray stated that *"embryology is to me by far the strongest single class of facts in favour of change of forms"*. Despite Darwin's use of embryology to support his views, Von Baer rejected the theory of common descent and did not regard embryological similarities as evidence for evolution.

Recapitulation Theory

The theory of recapitulation was a subsequent attempt to apply embryological similarity in support of Darwinism, which was first put forth in 1866 by Ernst Haeckel. He coined the phrase *ontogeny recapitulates phylogeny* to summarize the view.

- Ontogeny is the development of the organism.
- Phylogeny is the evolutionary history.

Haeckel suggested that *gill slits* and a *tail* could be seen during embryo development in organisms that did not possess them as adults. It was asserted that these organism must have possessed structures during their evolutionary history that were lost in adult stages due to natural selection. His theory held that during embryo development an organism progresses through stages in its evolutionary history-a theory that became known as the *Biogenetic law.*

It is important to note that the *gill slits* are not slits at all, but merely the folds in the region of the embryo's throat that are never involved with gas exchange (O_2/CO_2), which is accomplished by the umbilical cord. These folds, now called *pharyngeal arches,* later develop into the outer and middle ear as well as the bones, muscles, nerves, and glands of the neck. The appearance of an embryonic *tail* is simply due to the rapid growth of the spinal cord causing it to become longer than the body, and forcing the body to curl. This curling contributes to the folds visible in the neck region, in the same way a person develops a double chin when bending the neck forward.

Haeckel's Fraud

In an attempt to prove his theory true, Ernst Haeckel committed one of the most infamous examples of scientific fraud. In 1868, he published the series of embryos in the picture at right showing various living things placed side by side in the attempt to give the impression that there was significant similarity among them. The embryos shown in the earliest stages were altered to make them look more similar than they are in actuality. Embryologist Jonathan Wells notes that Haeckel was also very selective about the embryos he chose for his illustration. The four right-hand columns in the diagram at right are all from the same taxonomic subclass *Eutheria* (Placental mammals). Marsupials and monotremes were excluded, as were cartilaginous fish (sharks) and others. The fraud was exposed only months following the publication of these engravings by L. Rütimeyer, professor of zoology and comparative anatomy at the University of Basel, and corroborated by other contemporaries such as William His Sr, professor of anatomy at the University of Leipzig, who published their own comparisons showing significant differences (picture at left).

In 1997, the full extent of Haeckel's fraud was brought to light when Michael K. Richardson, a lecturer and embryologist at St George's Hospital Medical School, London, published photos of actual embryos illustrating the discrepancies between Haeckel's popular diagram and genuine embryos. Richardson described the situation in an interview in *Science* magazine as follows: *It looks like it's turning out to be one of the most famous fakes in biology.* The effort involved an international team of embryologists who examined and photographed *"the external form of embryos from a wide range of vertebrate species, at a stage comparable to that depicted by Haeckel"* because up until that time it was apparent that *"no one has cited any comparative data in support of the idea".* They analysed 39 different animals including those published by Haeckel and many that were not, such as marsupials, tree-frogs, snakes, and alligators, and they found significance differences with little conservation at the stage of development Haeckel claimed was most similar. In fact, the species in the fraudulent publication are so different that it is assumed the drawings made by Haeckel could not possibly have been done from real specimens.

Richardson summarizes the extent of their findings:

> *"We find that embryos at the tailbud stage – thought to correspond to a conserved stage – show variations in form due to allometry, heterochrony, and differences in body plan and somite number. These variations foreshadow important differences in adult body form. Contrary to recent claims that all vertebrate embryos pass through a stage when they are the same size, we find a greater than 10-fold variation in greatest length at the tailbud stage. Our survey seriously undermines the credibility of Haeckel's drawings, which depict not a conserved stage for vertebrates, but a stylised amniote embryo."*

In an interview for the London Times, Richardson is quoted as saying:

> *"This is one of the worst cases of scientific fraud. It's shocking to find that somebody one thought was a great scientist was deliberately misleading.... What he did was to take a human embryo and copy it, pretending that the salamander and the pig and all the others looked the same at the same stage of development. They don't ... These are fakes."*

Textbook Use

Despite the early exposure of this fraud by his contemporaries and following the almost universal recognition by the scientific community, Haeckel's embryo comparisons continue to appeared in textbooks in support of common descent. Examples of recent use include the 2004 edition of Biology: The Unity and Diversity of Life by Starr and Taggart., and the 1992 edition of Biology, a college level introductory biology text that states: *"In many cases the evolutionary history of an organism can be seen to unfold during its development, with the embryo exhibiting characteristics of the embryos of its ancestors. For example, early in their development, human embryos possess gill slits like a fish..."*

In 2000, the eminent evolutionist Stephen Jay Gould calls the continued use of these embryos in textbooks *atrocious*:

> *"We do, I think, have the right to be both astonished and ashamed by the century of mindless recycling that has led to the persistence of these drawings in a large number, if not a majority, of modern textbooks. "*

Homology

Similarities that exist between different organisms are cited by evolutionists as one of the most compelling lines of evidence for their theory. *Homologous* features are shared design concepts that are generally assumed by evolutionists to be the result of a shared ancestry and therefore offering support for the Darwinian theory of common descent. In The Origin of Species, Darwin referred to the *similar bones in the same relative positions* in the hand of a man and several other mammals as *specifically what the theory of common descent would expect.* The 2008 edition of Prentice Hall Biology by Miller and Levine illustrates that this concept is commonly held wherein it states: *In their early stages of development, chickens, turtles, and rats look similar, providing evidence that they shared a common ancestry.*

Much like gene comparisons, the similarities that are evident during embryo development (*von Baer's law*) are assumed outright to be due to phylogenetic relationships and therefore homologous. Furthermore, embryo homology is considered to be such powerful evidence of common ancestry that patterns during embryogenesis are used to determine if shared morphological structures are truly homologous (due to shared ancestry) or merely analogous (due to convergent evolution). Miller and Levine define homology in such terms: *structures that have different mature forms in different organisms but develop from the same embryonic tissues.* The reemerging of the early 19th century *"Von Baer's law"* has stimulated a new biological discipline called Evolutionary Developmental Biology, which is a collaboration of Evolutionary Biology and Developmental Biology, popularly known as Evo-devo. This area of scientific inquiry draw from the beliefs of Darwinism that all organisms have descended from the same ancestor, and generally assumes that embryonic developmental mechanisms should be shared (conserved). Drawing from these assumptions, evolutionary developmental biologists compare the developmental processes of different animals and plants in an attempt to determine the ancestral relationship between organisms.

Developmental biologist William Ballard a well known elder developmental biologist at Dartmouth College comments on the impact of *von Baer's law* on modern embryology.

"Thus, the energy of investigators and particularly students is diverted into the essentially fruitless 19th century activity of bending the facts of nature to support second-rate generalities of no predictive value. Though enthusiasm for Haeckel's (1900) recapitulation `law' died out, unfortunately the popularity of Von Baer's `laws' of 1828 was renewed. In order to defend the latter's descriptive statements that general characters appear before special characters as an egg develops and that the less general and finally the specific characters trail along later, we have to decide intuitively that certain characters are of `morphological significance' and others are not. When referring to vertebrates, we have to use words like blastula and gastrula in such a way as to imply that things that are vastly different from each other are really very much the same. "

Differences among organisms have traditionally been assumed to become more significant as development proceeds, and therefore earlier embryonic stages should resemble one another more than latter stages. This view can be found presented in popular biology textbooks. Although these assumptions have persisted into modern times, it has been recognized by embryologists since Darwin that this doctrine of developmental biology *"is not in accordance with the facts of development"*, according to British embryologist Adam Sedgwick (1894) who also stated: *there is no stage of development in which the unaided eye would fail to distinguish between them with ease"*. Modern embryologists confirm this as well, and have further challenged the erroneous nature of these assumptions. Recent work has revealed that embryos are not more similar during the earlier stages of cleavage and gastrulation, but in fact, are most alike at the later stage illustrated by Haeckel and recently photographed by Richardson. In the gastrulation stage, fish show dramatic differences when compared to amphibians, and neither are similar to reptiles, birds, or mammals.

Organisms begin life very dissimilar, then become somewhat alike during embryo development, then diverge again as adults. Evo-devos (evolutionary developmental biologists) refer to this pattern as the *developmental hourglass.* William Ballard adds that *only by semantic tricks and subjective selection of evidence can we claim that "gastrulas" of shark, salmon, frog, and bird are more alike than their adults.*

Ballard continues:

> *"In fact, the most obvious structural characteristics of either the eggs or the cleavage stages of a shark, a salmon, a frog, a bird, or a mammal are unique each to its own class, not generally shared. We would not consider them very much alike unless we had been taught so at a very early age. Very few vertebrates pass through a stage which can strictly be called a blastula. The embryo in its period of most active morphogenetic movements is usually called a gastrula, but as all agree this word has no morphologic meaning anymore. Each class of vertebrates (in mammals we might almost say each particular order) develops and then loses its own set of temporary structures-like the parade ground 'formations of maneuver' during this period. "*

Another significant challenge for evo-devo is the recognition that organs and structures develop from different groups of embryo cells in organisms that are believed to share a common ancestry. Fundamental structures, like the alimentary canal, kidney, or forelimb are commonly found to form from different embryological tissues. The alimentary canal (digestive tract) is formed from the roof of the embryonic gut cavity in sharks; whereas in frogs it is formed from the gut roof and floor; and in birds and reptiles it is derived from the lower layer of the embryonic disc or blastoderm. Vertebrate forelimbs, which were used as an example of homology by Darwin and can be found in hundreds of textbooks as proof for evolution, also develop from different body segments in different species. Rather than being proof of Darwinism, recent embryological studies have shown that forelimbs are not developmentally homologous. The forelimbs in the newt develop from embryo trunk segments 2 through 5; in the lizard they develop from segments 6 to 9; in humans they develop from segments 13 through 18. The development of the vertebrate kidney also conflicts with the view of homology. Evolutionist Michael Denton notes these conflicts in Evolution: A Theory in Crisis.

"In fish and amphibia the kidney is derived directly from an embryonic organ known as the mesonephros, while in reptiles and mammals the mesonephros degenerates towards the end of embryonic life and plays no role in the formation of the adult kidney, which is formed instead from a discrete spherical mass of mesodermal tissue, the metanephros, which develops quite independently from the mesonephros.

Studies of the genes governing development have added even further difficulties for evo-devo. According to the views of neodarwinists, homologous features are produced by genes that have been passed down by common ancestors. Because similar genes are unlikely to have originated independently through random mutation, sequence similarity between organisms indicates they share an ancestral relationship (they are homologous). Features produced by similar sequences are then also inferred to be phylogenetically homologous. For decades it has been recognized that homologous organs were frequently produced by different genes, but more recently, evo-devo biologists discovered that certain developmental genes were highly conserved (similar) in radically different organisms. These homeotic genes are short sequences that act as regulatory switches affecting the development of structures such as eyes and legs. When mutated, homeotic genes can transform one body part into another, such as an antenna into a leg. In organisms as vastly different as mammals and fruit flies, these genes are so similar that the mouse gene can regulate the development of the compound eye of the fruit fly. In such examples, the conflict with Darwinian assumptions lies in the fact that non-homologous structures are produced (or at least regulated) by supposedly homologous genes. As Jonathan Wells states very poignantly: *homology, whether defined morphologically or phylogenetically, cannot be attributed to similar developmental pathways any more than it can be attributed to similar genes.*

The famous embryologist Sir Gavin de Beer challenged his colleagues with the following question which remains unanswered to this day. "But if it is true that through the genetic code, genes code for enzymes that synthesize proteins which are responsible (in a manner still unknown in embryology) for the differentiation of the various parts in their normal manner, what mechanism can it be that results in the production of homologous organs, the same "patterns", in spite of their not being controlled by the same genes? I asked this question in 1938, and it has not been answered.

Worldviews

While Darwin was convinced that embryological homology was the *strongest single class of facts* to support his views. Modern embryology has instead challenged the Darwinian theory of common descent by illustrating that seemingly homologous structures, such as the mammalian forelimb, developed along different pathways, and are therefore not homologous at all.

While the Darwinists see these similarities as evidence of common descent, the creationist belief system causes us to generally conclude that shared structures, processes, or genes are merely reflections that they were designed by the same creator. Rather than being evidence of shared ancestry, homologies are examples of brilliant and well functioning designs that have been applied to multiple organisms, much in the same way as human designers apply concepts.

The Rise of Fish Embryology in the Nineteenth Century

It is particularly appropriate to write about the history of fish embryology at this time. History is often said to move in cycles, and this seems to be true of the study of fish development. After a period of prominence during the nineteenth and early twentieth centuries, the study of fish development, except in the hands of a stalwart few such as Oppenheimer and Trinkaus working with Fundulus and Ballard and Devillers working with the trout and other fishes, was overshadowed by developmental studies of other organisms, such as amphibians, the domestic fowl, sea urchins, and assorted invertebrates. For obvious reasons, the study of fish development continued to be pursued by fisheries biologists. With the popularization of the zebrafish Danio rerio as a "model of vertebrate development" and a revival of interest in the relationship between development and evolution, the cycle has come full turn and the study of fish development again has a place in the sun. My objective is to present an outline of the historical development of the science of fish embryology in the nineteenth century, broadly defined as 1789-1914. My plan is to assess the state of knowledge at the outset of the period, to document the advancement of descriptive embryology of teleosts and elasmobranchs, to consider the rise of comparative embryology of fishes associated with evolutionary studies, and to search out the beginning of experimental, physiological, and biochemical analyses of fish development.

When discussing the development of fishes, it is prudent to recall that fishes are the most diverse group of vertebrates. They should not be treated as a distinct, relatively homogeneous taxonomic unit. Rather, the term "fishes" applies to a grade of ectothermic, aquatic craniates and vertebrates and not to a distinct taxonomic group (Atz, 1985; Bruton, 1990; Nelson, 1994). Extant fishes are divided into five classes, namely Myxini (hagfish), Cephalaspidomorphi (lampreys), Chondrichthyes (cartilaginous fishes), Sarcopterygii, and Actinopteryg. Chondrichthyes contains two taxa, namely the subclass Elasmobranch (living sharks, rays, and skates) and the Holocephal (ratfishes or chimaeras) that diverged at a relatively early time, i.e., the Devonian-Carboniferous boundary. The class Sarcopterygii (flesh-finned fishes) contains the Actinistia (the living coelacanth), the Dipnoi (lungfishes) and the tetrapod vertebrates. The class Actinopteryg (ray-finned fishes) comprises five groups, namely Cladistia (bichir and reedfish), Chondrostei (sturgeons and paddlefish), Ginglymodi (gars), Halecomorphi (Amia calva, the bowfin), and the Teleostei (teleost fishes). Thus, the five classes of fishes contain eleven major taxonomic groups with an estimated total of 25,000 species. Each class is as distinct from the others as it is from the four other classes of vertebrates. The theme of eleven major taxonomic groups of fishes that is used in the ensuing presentation to emphasize the broad phylogenetic relationships of fishes and other vertebrates, tends, however, to obscure the extraordinary phylogenetic diversity within the teleosts. Teleostei, a monophyletic group, contains 24,000 species in 38 orders and 126 families compared to the 1,000 species, 19 orders, and 56 families in the ten other groups of fishes and the 25,000 species of tetrapods.

Not only are fishes the most diverse group of craniates, but they also display the greatest diversity of life history styles. Most fishes are oviparous, but viviparity is estimated to have independently evolved 42 times in five of the eleven major groups of fishes, all within the Chondrichthyes and Osteichythes. The phylogenetic diversity of fishes is reflected in the diversity and variability of their developmental patterns. Craniate development based on cleavage and subsequent patterns of gastrulation and embryogenesis is holoblastic, meroblastic, or transitional between the two. In holoblastic eggs, cleavage of the ooplasm is complete by the blastula stage. The morphogenetic movements of gastrulation occur in the three-dimensional space of the entire egg and involve most of the embryonic mass.

They often lead to the occlusion of the blastocoele. Neurulation is synchronous in time and space. Holoblastic development is considered the less specialized and more primitive craniate developmental pattern because it is characteristic of most echinoderms and non-craniate chordates as well as the more primitive representatives of various craniate lineages, namely lampreys, sturgeons, bichirs, lungfish, and most amphibians.

In the meroblastic egg, most of the yolk mass remains uncleaved at the blastula stage. Morphogenetic movements are confined to a portion of the egg volume, i.e., the blastoderm, and involve only a portion of the embryonic mass. Gastrulation and neurulation tend to be spatially and temporally asynchronous. Evolution of the meroblastic pattern appears to have involved qualitative and quantitative changes in the ooplasm and yolk components of the egg, namely an increase in total yolk content, an increase in the amount of yolk relative to ooplasm, a change in the type of yolk, and a change in the relative disposition of ooplasm and yolk. Not only did the changes in yolk-cytoplasm relationships of the egg affect cleavage patterns, but they also imposed a set of physical or physiological constraints on the morphogenetic movements of embryo formation, the so-called holoblastic-meroblastic barrier. Consequently, new or highly modified patterns of gastrulation and embryogenesis co-evolved with the evolution of the meroblastic egg.

The meroblastic pattern of development appears to have independently evolved in six craniate lineages, namely hagfish, sharks and chimaeras, teleost fishes, the living coelacanth (?), one group of amphibians (the caecilians) and amniotes. In each instance, different morphological patterns of gastrulation and embryogenesis have evolved. Thus, although meroblastic cleavage is superficially similar in hagfish, elasmobranchs, and teleosts, the morphological events of embryo formation differ considerably. These points and others are summarized in part by Collazo et al. (1994) who present a phylogenetic perspective and an evolutionary scenario for teleost gastrulation. Paradoxically, whether development proceeds according to a holoblastic or meroblastic pattern, the end product is a stereotypic craniate embryo. Recent advances in the study of the molecular basis of embryogenesis should help resolve this paradox.

Descriptive Embryology

Beginning with Aristotle, various authors have summarized the existing knowledge of fish development. In most instances, they did so in conjunction with larger works that dealt with the natural history or systematics of fishes, e.g., Rondelet (1554), Willughby and Ray (1686), Bloch (1782-1795) and Agassiz (1857), or else as part of a monograph that treated a selected aspect of development, e.g., Tilesius von Tilenau (1802), von Baer (1835), Duvernoy (1844), Van Bambeke (1876), Kingsley and Conn (1883), Henneguy (1888), and Scammon (1911). Comprehensive treatment of fish development was provided in several embryological textbooks, e.g., Balfour (1880), Ziegler (1902), and Kerr (1919).

Access to older, classical literature was provided by Dean's Bibliography of Fishes. It was not until the late Jane Oppenheimer (1936) published her "Historical Introduction to the Study of Teleostean Development" that the topic was subject to historical analysis. Only recently has the history of the study of elasmobranch development been summarized. Little historical documentation has been done for the other groups of fishes. When considering early studies of fish development, one soon realizes that many of these studies were carried out on viviparous species of elasmobranchs and teleosts rather than oviparous ones, and that both egglaying and livebearing elasmobranchs received a disproportionate amount of attention, considering their relatively low diversity compared to teleosts. First, elasmobranch eggs are considerably larger than those of most teleosts and their development can be easily observed with either the unaided eye or with a hand lens. Secondly, it has always been relatively easy to obtain gravid specimens of viviparous fishes as part of routine fishing efforts. Moreover, as museum or natural history collections of fishes were made during the course of explorations, specimens of curious, new species of gravid viviparous fishes became available for study, e.g., the four-eyed fish Anableps. In contrast, the eggs of oviparous teleosts are small and not easily collected in the wild. Artificial fertilization of teleost eggs did not begin until the late eighteenth century and the regulated spawning of captive fish species and the rearing of their eggs under "controlled" conditions for scientific purposes was first used only toward the mid-nineteenth century. In addition, the technology for maintaining fishes in aquaria and the possibility of spawning them under aquarium conditions did not emerge until the mid-late nineteenth century. In passing, one should recall that the Chinese have been rearing captive goldfish, including mutant strains, for over a thousand years. Other factors also influenced decisions to study certain viviparous instead of oviparous fishes. Viviparity in fishes, especially when there is a placental relationship, is an intrinsically interesting process. Finally, as a subject for the general study of development, oviparous teleosts have been secondary to the chicken egg ever since the Renaissance because of the greater size of the chicken's egg and its ready availability as an article of food.

Early History: Aristotle to Bloch

Because the research of early naturalists provided the foundation for the development of fish embryology in the nineteenth century, a brief outline of their work is presented here, beginning with Aristotle. Aristotle, the founder of the science of embryology, was the first to recognize and record many important features of fish reproduction and development (Aristotle, 1942, 1965, 1970). He distinguished oviparous (egg-laying) animals from viviparous (livebearing animals. Aristotle described viviparity in the dogfish shark (Mustelus), the thresher shark (Alopias), and the electric ray (Torpedo). He did not know of viviparity in teleost fishes because there are no viviparous teleosts in the coastal waters of the Mediterranean or the fresh waters of Europe and Asia Minor. His attribution of viviparity to the needlefish, a teleost, is actually a case of misinterpretation. The needlefish of Aristotle is a pipefish or sygnathid. Male pipefishes brood embryos in a pouch of abdominal skin. This form of parental care is analogous to viviparity and in its most highly evolved forms, the parentalembryonic relationships can be as complex as in viviparous fishes.

The body of Aristotle's observations concern chondrichthyan fishes and include: (1) the distinction between oviparous and viviparous modes of reproduction in sharks, rays, and skates; (2) description of the female reproductive system; (3) description of shark and skate egg cases; (4) description of egg structure and some observations on embryonic development; (5) first description of the oviducal or nidamental (=shell) gland but without any knowledge of its function in egg case formation or sperm storage; (6) hatching from the egg case in utero in viviparous species; (7) description of the male reproductive system; (8) notes on breeding seasons and inshore migration of gravid females for "pupping"; and (9) description of the yolk sac placenta in sharks. Writing of the smooth dogfish Mustelus canis in the 5th century B.C., Aristotle states, "the young are produced with the umbilical cord attached to the uterus so that as the substance of the egg gets used up, the embryo's condition appears to be similar to what is found in quadrupeds. The umbilical cord, which is long, is attached to the lower parts of the uterus; each one is, as it were, fastened to a cotyledon and is attached to the embryo by the middle where the liver is situated. Each embryo has a chorion and membranes of its own round it, just as in quadrupeds. If the embryos are cut open, a situation is disclosed exactly similar to quadrupeds; whatever internal organs they have, such as the liver, are large and supplied with blood".

Aristotle's observations on teleosts are in large part concerned with reproduction, e.g., his exquisite account of brood protection by male catfish. He realized that in contrast to the chondrichthyans, the overwhelming majority of teleosts are oviparous. His observations on the development of teleosts are more limited, due primarily to the small size of most teleostean eggs. Aristotle observed developing teleost eggs, recording their small size and rapid development. He states, "The embryo appears at the upper end of the egg and is enveloped in a membrane; the eyes, large and spherical, are the first organs visible. The embryo and the egg (yolk) are enveloped by a common membrane, and just under this is another membrane that envelops the egg (yolk) by itself; between the two membranes is a liquid". He described the organization of the teleost egg, distinguishing the perivitelline fluid from the yolk mass. [As an aside, Oppenheimer (1936) pointed out that some celebrated nineteenth century embryologists, viz. von Baer and Rathke, erred by considering the perivitelline fluid to be an albumen such as occurs in avian eggs.] Lastly, in comparing the development of teleostean and avian eggs, Aristotle recognized that teleosts lack the allantois.

Relatively little was added to the pioneering efforts of Aristotle by subsequent Greek and Roman scholars. With the decline of classical civilization, the study of science in general as well as the study of fish development languished for almost 1,000 years. Slowly the tide turned. According to Needham (1959), the first account of fish development in medieval Europe was given by Albertus Magnus, a thirteenth century scholastic. During the Renaissance, there was a revival of interest in natural history. The early zoological encyclopedists, especially Belon (1517-1564), Rondelet (1507-1566), and Aldrovandus (1522-1605), investigated fishes and described aspects of their reproduction and development. Again, there was a decided bias toward viviparous chondrichthyans. Rondelet (1554) illustrated an egg case and dissection of the female reproductive system of an oviparous catshark. He also illustrated part of the ovary and an egg case of a skate. He depicted a female shark, probably Mustelus canis, connected to a welldeveloped pup by an elongated yolk stalk that passes from the pup through the cloaca of the female. Although the placenta is not illustrated, Aristotle's research on the shark yolk sac placenta is cited in the text. Progressing to the seventeenth century, Fabricius (1600), who is best known for his studies of the chick, described and illustrated the female reproductive system and uterine embryos with large yolk sacs in a non-placental species of dogfish shark.

By 1673, Niclaus Steno had rediscovered the yolk sac placenta in the smooth dogfish, M. canis, investigated its anatomy, and published an illustrated account of it. He also distinguished between placental viviparity in the smooth dogfish shark, and aplacental viviparity in the spiny dogfish shark, Squalus. In the latter, he discovered that the yolk stalk terminates in an intestine characterized by a spiral valve. His rediscovery appears to have gone unappreciated. During this period, inability to distinguish among species of dogfish sharks that had either aplacental or placental modes of viviparity caused the very existence of a yolk sac placenta in sharks to be doubted. This controversy was not fully resolved until the time of Johannes Muller. Steno's studies on shark development had important ramifications for the study of mammalian reproduction. On the basis of his observation of large eggs in the uterine oviduct and large eggs in the ovary of the spiny dogfish, he concluded that the embryo does not form de novo in the oviduct, but comes form an egg released by one of the ovaries. From this, he postulated that mammalian development proceeded in a similar fashion, except that the mammalian egg is much smaller. In his examination of the mammalian ovary, unfortunately, he mistook the entire follicle for the mammalian egg. These studies were undertaken at a time when a controversy had arisen between animalculists and ovists with respect to the primacy of the sperm in reproduction and the absence of any knowledge of the mammalian egg. Steno's observations, although flawed, brought mammalian development in line with that of other vertebrates. In essence, he used sharks as a "model" system for mammals.

In 1678, Stephen Lorenzini, a pupil of Steno and Redi, published a set of observations on the viviparous electric ray Torpedo. This appears to be the first monographic treatment of a single fish species. Lorenzini's work is of considerable interest because almost one third of it is devoted to an account of urogenital anatomy, reproductive biology, and development of Torpedo, as well as several other rays.

Amongst his observations, those of primary interest concern oogenesis and ovulation, the blastoderm, and embryonic nutrition. Describing the ovary, he notes differences in size and structure between young and mature fishes. Maturing oocytes and ova vary in size, composition, and colour. Oocytes are connected to the ovarian matrix by a stalk through which several blood vessels pass, an arrangement similar to that described by Fabricius in the hen's egg. He states that these vessels provide nutrients to the eggs. The maturing egg is invested by a "membrane" except for the pole opposite the site of attachment to the stalk. This pole constitutes the pore or stigma through which the mature egg is released from the follicle. During oocyte maturation, changes in membrane adhesion were documented. Lorenzini speculates that mature eggs are released from the follicle by contraction of fibres in the ovarian stroma. The ovulated egg is "received" by the oviduct and passes into the uterine region where embryonic development is completed. Following Fabricius (1537-1619), De Graaf (1641-1673), and Harvey (15781657), Lorenzini refers to the residual portion of the follicle as the "calyx." Calyces, which apparently are the corpora lutea, occur in various sizes and shapes throughout gestation, and are lost or regress some time after parturition. Lorenzini found a correlation between the number of young in a brood and the number of calyces in the ovary. Ovulation in the Torpedo is considered to conform to De Graaf's account in other species. Among the more interesting of Lorenzini's observations are those made on the early stages of embryonic development. He describes and illustrates two neurula stages. The earlier stage depicts an early neural groove on the surface of the blastoderm. He describes the characteristic chondrichthyan blastoderm, "a leaden gray disc" that is clearly distinct from and surmounts the "yellow yolk mass." In another egg, he describes and depicts a slightly more advanced neurula stage (equivalent to Ballard et al., 1993, Stage 15). Here Lorenzini correctly shows the splayed, outspread cephalic margins of the medullary plate, a characteristic feature of chondrichthyan embryos. Lorenzini also observed the blastoderm in three other species of rays.

Lorenzini seems to be the first to have depicted the chondrichthyan blastoderm. His observations appeared only six years after Malphighi's initial account of the avian (chick) blastoderm (Needham, 1959). Lorenzini made one of the first major contributions to the knowledge of embryonic nutrition in non-placental viviparous fishes. Late stage embryos with yolk sacs lie free in the uterine region of the oviduct where they are bathed by uterine fluid. A yolk stalk passes from the external yolk through the abdominal wall and empties into a large internal yolk sac that in turn empties into the intestine.

The intestine is filled with yellow material apparently derived from the yolk and brown globules. The same brown globules occur in the uterine fluid and in the mouth, esophagus, and stomach of the embryos. Lorenzini concluded that advanced embryos are nourished both by yolk passed to the intestine and by uterine fluid which they imbibe. He believed that the nutrients in uterine fluid are derived from the blood and conveyed to the uterine fluid by vascularized glandular structures which he compared to chorionic villi. These structures are now known as trophonemata. Villi increase in size during gestation; their size is correlated with the amount of uterine fluid. They regress and are absent in non-gravid fish. In addition, Lorenzini also made significant observations on the male reproductive system, reproductive cycles, and the duration of gestation. In the closing years of the seventeenth century, Willughby and Ray (1686) summarized much of what was known of fish reproduction and development in a work primarily devoted to systematics.

Considering its promising beginning, it is surprising that the study of chondrichthyan development made relatively little progress in the eighteenth century. There are many possible reasons. Those scholars interested in fish biology may have been attracted to systematic studies because of the revolutionary efforts of Willughby and Ray, Artedi (1705-1735), and Linnaeus (17071778) as well as the influx of new species of fishes from the tropics. Changing fashions in science also enter the picture, especially the rise of microscopy and microscopic anatomy. Perhaps in terms of the material, questions, and technology, research on chondrichthyans had reached an impasse.

There are some exceptions. Bohadsch (1776) investigated the structural organization of eggcases. Monro (1785) published detailed descriptions accompanied by elegant illustrations of the urogenital anatomy of the oviparous skates. He also described and illustrated for the first time the spermocysts, spherical aggregations of Sertoli cells and differentiating sperm cells that are a characteristic feature of chondrichthyan spermatogenesis. Descriptions and illustrations of microdissections of the internal anatomy of mid-and late-stage embryos are given. The extensive vitelline circulation and the connection of the yolk sac and the small intestine are documented. External gill filaments, another feature of chondrichthyan development, are described for the first time. The eighteenth century concluded with Bloch's magnum opus (1782-95) which added little to the study of chondrichthyan development other than information on the embryos of the viviparous sawfish Pristis. The study of teleostean development, as previously noted, lagged well behind that of both oviparous and viviparous chondrichthyan fishes, and little was done prior to the mid-late eighteenth century. To be more accurate, the little that was done was done with viviparous teleosts. Schoenveld (1624) first described viviparity in the embryos of the eelpout, Zoarces viviparus, a fish common to the marine waters of northern Europe. Late stage embryos are illustrated. Impressed by the substantial growth of the embryos during gestation, Schoenveld postulated that they obtain nutrients by ingesting ovarian fluid. (In viviparous teleosts, an oviduct or uterus is absent and gestation takes place in the ovary. Cf. Wourms, et al., 1988). Somewhat later, Willughby and Ray (1686) apparently observed the development of live Zoarces embryos. Interest in teleost development during the seventeenth and eighteenth centuries seems to have been generated by the viviparity of exotic species discovered during the explorations of the Old and New World Tropics. Anableps, the four-eyed fish, conspicuous because of its size, unusual anatomy, and behaviour in its Central and South American habitats, was described as viviparous as early as 1738. Gronovius (1754-1756) was the first to illustrate advanced embryos of Anableps with their characteristic rugose "yolk sac" (actually the abdominal trophoderm). Gronovius may also have been the first to describe viviparity in the clinid, Clinus superciliosus. Bloch (1782-1795) described and illustrated the external and internal reproductive anatomy of Anableps, its embryos in situ, and the embryos themselves. He also depicted the adults and embryos of Zoarces and Clinus. One of the first sets of field observations was made by de Alzate — Ramyrez in 1769 who discovered viviparity in a Mexican poeciliid and made microscopic observations of blood circulation in living, near-term embryos.

During this period, the foundations were laid for the study of early development, i.e., fertilization through embryogenesis. In 1763, Jacobi (cited in Oppenheimer, 1936) published an extensive letter in which he reported that he had artificially fertilized the eggs of trout and salmon by removing the ripe eggs and adding sperm to them. Although this is the first known written account of artificial fertilization, it is probable that the technique had been in use for some time, for fish culture had been practiced in Europe since the late fifteenth century. Jacobi may have observed the entry of sperm into the egg. He also recorded the formation of double embryos in artificially fertilized eggs, a condition that he attributed to the entry of two spermatozoa. In 1787, Cavolini published a monograph on the "generation" of fishes and crabs. The bulk of this work was devoted to the description of spawning habits, hermaphroditism, fertilization, and contemplations on the nature of development. However, he did make one observation of considerable significance. In the eggs of the pipefish, he described a structure on the egg which he referred to as the "cicatrix" or in modern terminology, the blastoderm.

He identified the blastoderm with the same structure in the chick and assumed that the embryo originated from the blastoderm. The study of teleost development in the eighteenth century was brought to a close with the appearance of Bloch's (1782-1795) massive, twelve-volume monograph on the natural history of fishes. This work is one of the linchpins of systematic ichthyology and summarized what was known about fishes up to that time. It contains much information on fish reproduction and development. As previously noted, aspects of the development of viviparous chondrichthyans and teleosts are described and illustrated. Of greater interest are his pioneering studies of the early stages of development of an oviparous teleost. For the first time, he describes and illustrates consecutive stages in development, from post fertilization through hatching to a late larval stage, and provides a time base for the description of developmental stages. He appreciated the eccentric position of the yolk mass relative to the blastoderm. He made observations on the development of the heart, and illustrates the contractile, simple tubular stage of heart development. There were, of course, errors of interpretation, one of which was the identification of the perivitelline fluid of the teleost egg with the albumen layer of the avian egg.

Development of Teleostean Fishes

As the nineteenth century dawned, the foundations had been laid for the study of fish development, and the stage had been set for the extraordinary progress that would soon be made. The rise of fish embryology was vitally linked to the origins of embryology as a formal science and its subsequent growth, diversification, and interaction with other branches of biology. Scientific embryology is generally considered to begin with the publication of the first volume of von Baer's comparative studies of animal development. As Oppenheimer (1936) has pointed out, even though von Baer's work codified embryology as a science and profoundly influenced its future directions, nonetheless, it was the culmination of a course that had been pioneered by Dollinger (1770-1841) and Rathke (1793-1860). Oppenheimer (1936) states that Dollinger "was the originator of the appreciation of comparative development by working out carefully the succession of steps in the developmental processes in various vertebrate forms." Dollinger not only influenced von Baer (1792-1876) but also guided Pander's (1793-1860) revolutionary studies of chick development and inspired in Louis Agassiz (1807-1873) a life-long commitment to a program of comparative embryological research. Dollinger's approach to embryology had its origins in the influence of German Naturphilosophie on the study of comparative anatomy. Thus influenced, anatomists attempted to systematize, describe, and classify all organisms or parts of organisms as variations of a single type. Dollinger, who was a proponent of Naturphilosophie, thus accepted and extended suggestions of the anatomist Meckel (1781-1833) that a unity of plan manifests itself in development as well as in structure.

For reasons of convenience and also because their study progressed at different rates and in different ways, I have chosen to treat teleosts and chondrichthyans separately. In retrospect, it is amazing to contemplate the almost explosive increase in knowledge of teleostean development that took place in the period 1819-1842. The first major work from this period was Forchhammer's (1819) dissertation on the development of the viviparous blenny Zoarces. He described no stages younger than pre-motile embryos. The main body of the work comprises descriptions of older embryos and organ formation. He was the first to describe the embryonic shield, the region in which the embryo forms. His observations on the contractility of the excised embryonic heart constitute one of the earliest attempts to study isolated embryonic organs. He also observed that yolk ultimately is absorbed into the liver. Rathke (1833) also published a monograph on the development of the viviparous blenny, as part of a much larger serial work on the development of animals and humans. Conceptually, Rathke's studies parallel those of Dollinger and von Baer (Oppenheimer, 1936). Although he described and analysed early stages of development, he was primarily concerned with older stages. His observations on the developmental anatomy of the brain, heart, gut, and other organs established a new standard of excellence in the field.

The first volume of von Baer's three-volume treatise on embryology appeared in 1828 but fishes were not dealt with until the second volume. In the interim, he published a separate communication devoted to the development of fishes based on his studies of species of Cyprinus. He made several fundamental observations such as the sharp distinction between the pattern of development in large-yolked and small-yolked teleostean eggs and the great differences in the duration of development, viz. three months in trout, a few weeks in Zoarces and a few days in Cyprinus. His choice of Cyprinus, an egg-laying species, made it possible to obtain newly fertilized eggs, thus giving him access to the early, i.e., cleavage-epiboly, stages of development and providing a chronological base for the consecutive stages of development. von Baer (1835) was the first to describe clearly and illustrate one of the most characteristic features of teleostean development, viz. epiboly or overgrowth of the yolk by the blastoderm. Not only did he describe it, but also he provided a timetable for the process. He was also the first to describe the germring, i.e., the thickened edge of the blastoderm. von Baer failed to observe cleavage and incorrectly homologized teleost development with that of other vertebrates by mis-identifying the early embryonic shield as primitive streak and the late shield as a broad neural plate. von Baer concluded his monograph on fish development with an appendix in which he discussed the embryological origin and function of the swimbladder and compared it to the lungs of other vertebrates.

Studies very much similar to those of von Baer were being carried out by Rusconi (1836) on two species of Cyprinus. Rosconi gives a detailed account of his method of artificial fertilization. Developing eggs, thus obtained, were of known age and were subsequently examined microscopically. His major contribution is that he appears to have been the first to describe cleavage in the teleostean egg. His detailed description is illustrated by a plate which "rivals all other in the literature for accuracy and beauty in presenting the early stages of teleostean development".

He also made extremely accurate observations on the embryonic shield and its role in the formation of the embryo. To aid observation, he used an acid to stop development and render the embryo opaque without affecting the transparency of the yolk. His figures depict the concentration of cells in the midline of the embryonic shield that are the precursor of notochord and spinal cord. Elsewhere, he states for the first time that the teleostean nerve cord originates from a single structure rather than by the folding of a neural plate.

Embryological research on teleosts moved to France during the middle years of the century. The first important work was Vogt's (1842) contribution to the embryology of the whitefish (Coregonus), which was part of Agassiz's elaborate series of monographs on the freshwater fishes of central Europe. The embryology portion consists of a text volume and a folio of coloured lithograph plates. Although considered part of the French School by Oppenheimer (1936), it could just as well be considered a product of the Dollinger-Wurzburg School.

Agassiz was in close contact with Dollinger at Munich between 1827 and 1830 and was studying teleost development. The role of Agassiz, if any, in Vogt's study is not clear, for the dispute over authorship was the first in a series of controversies that would occur between Agassiz and his collaborators. In addition to its scientific methods (vide infra), this work had an extraordinary impact because of the clarity and beauty of its illustrations. The figures were drawn by professional artists and reproduced, many in colour, by lithography, a printing process capable of rendering fine detail. Not only were new observations effectively presented, but also the known details of development took on a new lucidity. The Vogt-Agassiz study is another study in which developing eggs were obtained by artificial fertilization, cultured within a set temperature range (5-IOC), and examined microscopically. For the first time, the technique of dechorionation is mentioned as an aid for observation. Vogt also used the technique of acid fixation to enhance the visibility of the embryo. A series of well-illustrated, closely spaced stages in the normal development of the whitefish was presented. This work appears to be the first study to apply a cellular approach to the interpretation of developmental events, especially oogenesis, cleavage, epiboly, and embryogenesis. Information is provided on oogenesis, the structure of the mature egg, and fertilization. According to Oppenheimer (1936), Vogt was the first to inquire into the mechanics of embryo formation. He also speculated on the role of cell movement and cell migration in development. A considerable portion of the work was devoted to the microanatomy of the later stages of embryonic and larval development. In this respect, Vogt gives an account of the formation of a number of organ systems, tracing their development from the earliest cellular condensations or anlagen to completion. Illustrations of melanocytes and the cellular components of the notochord and connective tissue are remarkable for their detail which included nuclei.

The next figure of consequence in the French school was Lereboullet, a pupil of Duvernoy, who in turn was a protege of Cuvier. Lereboullet made substantial contributions in the field of descriptive embryology and subsequently in experimental studies which were the first great attempt to analyse the mechanism of embryo formation. His monograph on the development of the pike and perch established new standards in terms of the accuracy with which his observations were depicted. He provided the first detailed description of the microscopic organization of the egg and a more accurate description and illustration of the embryonic shield. He stated that it is the precursor of the entire embryo and not just the central nervous system. His most important single observation was the description of the yolk syncytial layer (periblast) which he termed "feuillet muqueux," literally mucous lamina. He mistakenly ascribed to it the fate of forming elements of the digestive system. He regarded the germ ring, i.e., the thickened margin of the blastoderm, as a dynamic force during the formation of the embryo. This viewpoint, reinforced by his studies on experimental teratology (vide infra), led him to interpret the process of embryo formation in terms of concrescence, i.e., the right and left portions of the germ ring assemble into the two halves of the body. Lereboullet arrived at this conclusion a decade before His (1874) formally promulgated the concrescence theory. Finally, in his monograph on the comparative embryology of the trout, a lizard and a snail, he refuted the concept of a unity in the plan of development propounded by Meckel and Serres.

Three important structures, characteristic of teleostean development, still need to be dealt with, namely the yolk syncytial layer or periblast, the enveloping layer, and Kupffer's vesicle. Although Lereboullet (1862) first described the yolk syncytial layer, it was von Kupffer (1868) who suggested that the cells of the blastoderm not only give rise to the tissue cells of the embryo but also are progenitors of the yolk syncytial layer. Agassiz and Whitman (1884, 1915) proved beyond any doubt that yolk syncytial nuclei are derived from the marginal cells of the blastoderm. The enveloping cell layer is a tightly sealed epithelial monolayer that forms the outermost boundary of the blastoderm. Priority for its discovery is unclear, for descriptions of it under various names appeared more or less simultaneously in a number of independent studies. Kupffer's vesicle is a transient structure that is only found in teleost embryos. It was discovered by von Kupffer (1866), who thought that it might be an allantoic rudiment. It is a small, fluid-filled epithelial sac that appears during the somite stages. Kupffer's vesicle develops mid-ventrally in a region below the tail bud and at the interface between the embryo and yolk mass. Even now, its function, if any, still remains unknown.

At this point, a digression to consider the development of viviparous teleosts is in order. Following the pioneering studies of Forchhammer (1819) and Rathke (1833) on Zoarces, knowledge of viviparity in teleosts progressed steadily through the rest of the century. In 1844, Duvernoy described the development of a freshwater poeciliid fish from Surinam. His work was mainly concerned with the later stages of development, but did provide some important information on early development. The ovary is the site of gestation. Fertilization takes place in the follicle where the embryo is invested by an egg envelope. Development to term takes place in the follicle. Duvernoy illustrated for the first time an early stage follicular embryo in its egg envelope. In his Memorias, published 1851-55, the Cuban naturalist Poey was able to extend the number of viviparous species to include all ten species of Cuban poeciliids. Of greater significance were his observations for the first time of sperm storage in these fishes. Unfortunately, he interpreted his observations in terms of parthenogenesis rather than sperm storage and superfetation. The problem was not satisfactorily resolved until 1908, when Phillipi discovered that in poeciliids, sperm is packaged into spermatozeugmata that are transferred to the male reproductive tract. He also presented evidence for specific structural pathways, which he called "delle," that allow sperm to move from the lumen into close proximity with the egg and fertilize it within the follicle. In 1856, Wyman produced a definitive study on the embryology of the foureyed fish Anableps. It, too, is a fish with follicular gestation. He depicts the elaborate follicular epithelium and the highly vascularized abdominal trophoderm of the embryo. Because of the massive increase in size of the embryo during gestation, he postulated that the developing embryos are nourished by secreted maternal proteins. He extended his speculations to include embryonic nutrition in the surfperch and the electric ray Torpedo.

In 1853, Agassiz first described viviparity in the surfperches or embiotocids from the Pacific coast of North America. In a more extensive publication, Girard (1858) described and illustrated 17 species. He provided the first detailed information about their reproduction and development. Gestation is ovarian: embryos develop in the ovarian lumen and are enfolded by sheaths of ovarian tissue. Initial egg size is minute, but term embryos are large (5 cm). A series of developmental stages is described and illustrated. Figures on brood size which varies (10-80 young) according to species are given. The duration of gestation is estimated at 5-6 months. The initial descriptions of Agassiz and Girard culminated in a definitive study of surfperch development by Eigenmann (1894) which is tour de force of descriptive embryology. He noted that the early cleavage stages of the small, yolkpoor eggs verge on being holoblastic. He described the presence of sperm in the ovarian fluid and the imbibition of ovarian fluid by the embryos. Male embryos undergo precocious sexual development. In an earlier separate publication, he described the precocious segregation of germ cell determinants at an early cleavage stage, an overlooked, early contribution to concepts of germ cell development. Throughout the remaining years of the century the occurrence of viviparity was described in various groups of teleost, e.g., the rockfish, Sebastes, halfbeaks, Hemirhamphus, and the Lake Baikal oilfish, Comephorus. One should recall that such studies are still continuing, with viviparity in the living coelacanth Latimeria not discovered until 1975. By 1886, the field had progressed to the extent that Ryder (1886a) could review it and add to it some of his own original observations. He concluded that the developing embryos of surf perches derive nutrients by absorbing imbibed ovarian fluid through elongated intestinal villi found in the hypertrophied hindgut. He also concluded that the other characteristic feature of surfperch embryos, the unique vascular apparatus of the enlarged vertical fins, serves a respiratory rather than a nutritive function. Much of the research on viviparity, both in teleosts and chondrichthyans during the latter part of the century was concerned with the elaboration of highly diverse maternal and embryonic structural adaptations for viviparity and attempts to correlate these structures with physiological function, e.g., nutrition, gas exchange.

Advances in technology had profound effects on embryological research during the latter half of the century. Because of the transparency of most teleostean eggs observations had usually been made on intact eggs or microdissections of embryos. Beginning in 1860, a series of discoveries and innovations in histological microtechnique made it possible to fix, embed, section, stain, and subsequently examine serial sections of embryonic material with microscopes equipped with high resolution, apochromatic, oil-immersion objectives (cf. Ryder, 1884b and Whitman, 1885 for the availability of techniques in embryology, and Horder et al., 1985, for a chronology of technological and conceptual innovations). In this respect, the application of photography to the study of fish development occurred surprisingly early in the history of photography. Von Kupffer is credited with the first extensive use of photomicrographs to illustrate an embryological monograph in his 1878 study of herring development. By 1898, His had published micrographs of sectioned trout blastomeres that illustrated aspects of chromosome behaviour during cell division.

The consequences of these advances in embryological technology was to trigger a veritable explosion of embryological studies of teleosts based on the microscopic study of sectioned material, e.g., Van Bambeke, 1876; Hoffman, 1881; Henneguy, 1888; Ryder, 1884a; Kowalewsky, 1886; and Wilson, 1891. By the end of the nineteenth century, an excellent descriptive knowledge of teleost development had been achieved. But there was a trade-off. The emphasis on the study of sectioned material detracted from the observation of living embryos, especially cell behaviour during early development, a field that was just beginning to be explored by His (vide infra). This trend led to subsequent difficulties in the interpretation of morphogenetic movements and embryo formation. Even as this trend was established, its advantages and limitations were becoming obvious. Harrison (1901) published his histological study of the differentiation and axonal growth of spinal neurons in trout embryos in which he focused on the patterns of growth of individual nerve cells and their role in establishing neural connections. The constraints of the histological approach led him to develop tissue culture and use it six years later to investigate neuronal outgrowth.

Development of Chondrichthyan Fishes

At the onset of the nineteenth century, the study of chondrichthyan development was already farther advanced than that of teleosts. However, the very features that had initially favoured the study of chondrichthyan fishes, e.g., large eggs and embryos, viviparity, and internal fertilization, subsequently worked against their further study. It was difficult to obtain access to early stages of development in viviparous species and there was no really effective way to maintain egg-laying species in captivity so as to obtain newly laid eggs. The rate of development was slow. The large, meroblastic eggs were opaque, so that only external features of development and those details revealed by microdissections of embryos could be observed. In contrast, the introduction of artificial fertilization made the early stages of teleostean development readily available and the transparency of most teleostean eggs and embryos was well suited for the microscopic examination of external and internal features of development. Accordingly, nineteenth century studies of chondrichthyan development can be divided into two phases. Phase One spanned the period 1800-1860. During this phase, research progressed along previously established lines with an emphasis on reproductive biology, the study of viviparous species and adaptations for viviparity, and the later stages of embryonic development. Phase Two spanned the period 1860-1911. Research in this phase was shaped by the introduction of new embryological and histological microtechniques to the study of chondrichthyans.

Research of the eighteenth century passed imperceptibly into the nineteenth century with little or no change of agenda. In addition to his magnum opus, Bloch published a set of separate papers in obscure journals, some of which dealt with chondrichthyan reproduction and development. Tilesius von Tilenau (1802) published a monograph on the egg cases of sharks and skates. It is noteworthy because it represents the first attempts to determine the chemical composition of egg cases, now known to be a type of "tanned" protein. In addition, he also reported on the urogenital anatomy and the reproduction and development of sharks and skates. The major significance of Mitchell's (1803) report on the anatomy of early (7-8 cm) pre-implantation shark embryos is that it is one of the earliest embryological papers and probably the first paper on fish embryology published in the United States.

Home's (1810) paper had two major aspects: one embryological and the other reproductive. He provided anatomical descriptions of the male and female reproductive system and described the structural changes that occur as mature individuals pass from a reproductively inactive to an active state. His study of the male urogenital system and claspers (modified caudal portion of each pelvic fin) represents the first step toward understanding the process of chondrichthyan insemination and fertilization. He gives a fine illustration of the claspers with their terminal processes expanded as they would be in copulation. He also describes the urogenital papilla, the terminal end of the male reproductive duct, but mistakenly refers to it as a "penis" and states that it is the intromittent organ.

Claspers were first reported by Aristotle, and their gross anatomy described by Lorenzini (1678) and Bloch (1782-1795). A controversy developed over the centuries concerning their function either as an intromittent organ or an organ for holding the female fast during copulation Further contributions were made by Davy (1839). He described the internal anatomy of the clasper that makes possible its function in sperm transfer. On the basis of his own work and a review of the literature, he concluded that the clasper functions both as a holdfast and intromittent organ. He also described the accessory clasper gland of skates and rays, a structure now known to produce prostaglandins (Lacy, personal communication). Davy (1839) appears to be the first to have observed chondrichthyan sperm. In the electric ray (Torpedo) and a skate, he described sperm with an elongated head and flagellum. In vitro microscopic examination revealed that their motion was "serpentine and vibratory." Details on the process of spermatogenesis and the structure of the mature sperm came later. In 1871, Agassiz stated that when the claspers (of a skate) are rotated forward and upwards, an opening in them was brought opposite the spermatic duct (=urogenital papilla). The clasper could then be introduced into the oviduct and spermatic fluid could pass up the clasper and into the oviduct. Bolau (1881) appears to be the first to actually observe and describe the mating activity of sharks. The current view of insemination is that of Gilbert and Heath: "In mating, one clasper is flexed medially, inserted, and is anchored in the oviduct by a complex of cartilages at the clasper tip. Sperm pass from the urogenital papilla into the clasper groove and are washed into the oviduct by seawater and secretions expressed from a siphon sac." The actual cytological events of fertilization in chondrichthyans did not become known until Ruckert (1892b) demonstrated physiological polyspermy.

Embryology was the second aspect of Home's (1810) paper. He compared stages in development of the oviparous catshark, Scyliorhinus with the non-placental viviparous spiny dogfish shark, Squalus. He illustrated an advanced yolk sac embryo of the catshark in its egg case. He illustrated and described the slits that open in the egg case to permit the flow of seawater through the egg case during mid-late phases of development. He gave an excellent rendering of an oviduct of gravid Squalus with its ostium tubae, nidamental (shell) gland, and posterior uterus lined with longitudinal vascular ridges. Within the uterus, the "candle" is illustrated for the first time. It is an attenuated egg case that contains three developing eggs and a quantity of transparent jelly. Neonates of both the catshark and spiny dogfish are illustrated. To determine the physical and chemical properties of the oviducal jelly of the spiny dogfish, Home (1810) enlisted the aid of the chemist W. Brande. The jelly was extremely hydrophilic, being able to absorb many times its volume of water. Chemically the jelly was found to be neither gelatin nor albumen. It is interesting to note that Home, under the influence of Humphrey Davy, speculated on the requirements of developing fish eggs for aeration and how this process might be accomplished with different patterns of embryonic circulation.

Rathke was von Baer's successor at Konigsberg. His study of the "sexual apparatus" in fishes and other vertebrates was the first major step in understanding the development of the urogenital system, the differentiation of the gonads and accessory ducts, and the clarification of relationships in the vertebrate urogenital system. His monograph on the development of sharks and rays was primarily concerned with organogenesis and the microanatomy of the developing embryo. His comparative study of the development of the gill slits and arches forms the basis for homologizing the pharyngula phase of vertebrate development and was used extensively by the proponents of the biogenetic law later in the century. John Davy, the brother of Humphrey Davy, published an account of the embryonic development of the electric ray Torpedo and a series of experiments on its electric organs. In his description of the female reproductive system, he noted that the absence of a nidamental (shell) gland accounts for the absence of an egg envelope in this fish. He presents an illustrated series of developmental stages which includes information on the electric organ. His earliest embryo was 18 mm long. It had short, external gill filaments but lacked an electric organ. The electric organ had begun its development in 28 mm embryos. Pectoral fin development began in 39-35 mm embryos, and by 62 mm the embryo looked like a miniature adult. Living full-term embryos were surgically removed from the female and maintained in sea water for up to six months. These neonates had functional electric organs. Activity was measured with a galvanometer. At six months, the young rays still had vestiges of an internal yolk sac and showed no interest in food. Davy suggests that they depend on yolk reserves during this period. Davy's (1834) contributions to the study of embryonic nutrition will be discussed in another section.

Almost simultaneously, Leuckart (1836), a protege of Rathke, published a comparative study of the external gill filaments of several species of sharks and the electric ray. This paper has added interest because Leuckart's incorporation of an extensive series of measurements of embryonic structures represents an early attempt to quantify embryological research. Based on his own work and that of others, Leuckart concluded that external gill filaments are transitory structures that are common to the embryos of all chondrichthyan fishes. He suggested that passage of the embryos through an external gill filament phase is a type of metamorphosis. In addition, for the first time he describes the yolk stalk appendiculae of the pre-implantation embryos of the bonnethead shark, Sphyrna tiburo. They are vascularized villiform projections that festoon the umbilical cord of this placental species.

Johannes Muller was one of the pivotal figures in mid-nineteenth century biology as a pioneer marine biologist, a founder of comparative physiology, an accomplished embryologist, and a mentor of illustrious students such as Henle, K6lliker, Haeckel, His, Virchow, and Du Bois-Reymond. He helped to lay the foundations for the experimental approach to embryology. Muller's classical paper of 1842 is considered a benchmark of chondrichthyan research. In it, he employed a comparative approach to the study of oviparous and viviparous species. He described and compared the egg cases of sharks, skates, and chimaeras. More importantly, he provided a detailed description of the yolk sac placenta in the smooth dogfish shark, Mustelus canis and the blue shark, Prionace glauca. He gave an in-depth account of the maternal and fetal portions of the placenta and their vascularization.

Prior to Mller's definitive account, knowledge of the shark yolk sac placenta had languished since Steno's (1673) rediscovery of it, over 170 years previously. Shortly thereafter, Leydig's (1852) monograph on the histology and embryology of rays and sharks appeared. In it he described five stages in the early development of the spiny dogfish, Squalus acanthias. A detailed description and accurate illustration of the microanatomy of the earliest stage (equivalent to Stage 23 of Ballard et al., 1993) is given. His research on the yolk sac placenta extended that of Miller. The embryonic portion of the placenta of M. laevis at its attachment site is illustrated and a schematic illustration of the histology of the maternal embryonic placental interface is provided. In another advance, Leydig (1852) described the differentiation of six tissues, viz. neurons, notochord, connective tissue, lens fibres, heart muscle, and striated muscle. In the latter tissue, cell fusion was depicted.

Although the eggs and embryos of viviparous species continued to be used for the study of development, the study of viviparity per se entered a period of stasis that lasted well into the twentieth century. However, some progress was made. Bruch (1860) published the first accurate descriptions and illustrations of the uterine villi of several viviparous rays. The extensive vascularization and glandular organization of these structures were noted. These observations were confirmed and extended by Trois (1876). Wood-Mason and Alcock (1891) discovered that the uterine villi of the butterfly ray (Gymnura) are grossly hypertrophied in one region of the uterus and enter the spiracles of the embryo. This juxtaposition of maternal and embryonic tissues was subsequently termed "a branchial placenta". WoodMason and Alcock coined the term "trophonemata," literally growth threads, to describe the hypertrophied uterine villi. They ascribe a function of nutrient production to the trophonemata. Developing embryos would imbibe the nutrients. Although studies of the yolk sac placenta also lagged, Ercolani's (1879) observations on shark placentae were incorporated into his general classification of the types of vertebrate placentae. Parker (1889) described the fetal membranes of Mustelus antarcticus and provided an analysis of the periembryonic fluid. Alcock (1890) described the histology of the spatulate extensions that occur on the umbilical cord of the yolk sac placenta in the hammerhead shark Sphyrna blochi. He uses, apparently for the first time, the term "appendiculae" to categorize these processes. Their function was enigmatic. Alcock speculates on a possible role as a lymphatic gland or involvement in the "purification" of embryonic blood.

Advances in embryological microtechnique and microscopy that so profoundly altered the study of teleostean embryology had an even more pronounced effect on the study of chondrichthyans because of their large, opaque, meroblastic eggs. Research shifted to the histological and cellular levels. The consequences are especially apparent in research on early development, i.e., fertilization through early development, as well as organogenesis and cell differentiation.

Although the blastoderm and neurula stages had been recognized by Lorenzini (1678), little progress had been made in the intervening years in defining the blastula and gastrula stages. Coste (1850) appears to be the first to describe but not illustrate cleavage. His report was extended by Gerbe (1872) who illustrated surface views of stages from the first cleavage through a late blastula and compared them to the cleavage stages of birds and reptiles. These observations were confirmed by Balfour (1878) who also included illustrations of sectioned blastulae. The definitive study of early development from fertilization through the early gastrula was conducted by Ruckert (1899). Surface views of the blastoderm were accompanied by sections through it.

In 1878, Balfour published his monograph, On the Development of Elasmobranch Fishes. This work raised the study of chondrichthyan development to heretofore unattained heights by establishing new standards of excellence for accuracy and comprehensive description. It has been categorized as "a famous landmark in the history of vertebrate development". Balfour explored all aspects of elasmobranch embryology from early cleavage through mid-late development and established a staged series of embryos, mostly sharks, that was employed in research for many years. Illustrations of sectioned material were used extensively. Balfour described the organization of the mature egg and cleavage of the zygote, commenting on yolk platelets, the germinal vesicle and its fate, the process of cleavage, nuclear division, the origin of yolk nuclei, and asymmetry of the blastoderm. An indepth account of gastrulation, germ layer formation, and the general details of embryonic development is presented. Detailed descriptions of the development of the major organ systems included: (1) brain, spinal and sympathetic nerves; (2) head and associated structures; (3) digestive tube and associated organs; (4) heart and vascular system; and (5) gonads, kidney, Wolffian and Mllerian ducts and their derivatives. In passing, it should be noted that Balfour conducted much of his research at the Naples Zoological Station. His research was facilitated by the director, Anton Dohrn, who was pursuing a major research program in evolutionary physiological anatomy. This program included studies of shark embryology by Dohrn and his associates in order to resolve Dohrn's ideas about vertebrate origins.

Observations by Ziegler and Ziegler (1892) on the development of the electric ray Torpedo continued Balfour's pioneering approach. Their work is characterized by the use of serial sections and photographs of surface views of model reconstructions of selected embryonic stages.

They investigated the equivalent of Balfour stages B to K, and considerably advanced descriptive knowledge of gastrulation, formation of the medullary plate, closure of the medullary plate and early embryo formation, and the organization of gill slit-tail bud stage embryos. Balfour had made major advances in the study of chondrichthyan development and the introduction of a series of normal stages was useful. However, there were defects in his normal series. The series was intended to illustrate development in Scyliorhinus, but it was necessary to use embryos of the shark Galeus and the ray Torpedo to fill in gaps. Moreover, there was no time base to the series. As part of Keibel's series of monographs on the normal stages of development of vertebrates, Scammon (1911) published his monograph on the spiny dogfish shark, Squalus acanthias. It is a reasonably complete survey of the development of a single elasmobranch species. Ballard et aL (1993, p. 334) refer to Scammon's monographs as "hitherto unrivaled in its completeness, records in handsome drawings, wax reconstructions and serial sections the anatomy of dozens of randomly collected but closely spaced specimens." Scammon (1911) also included a bibliography arranged by subject of the important literature on chondrichthyan development. The appearance of Scammon's monograph represents the apex of nineteenth century studies of chondrichthyans.

A major conceptual difficulty encountered in the study of chondrichthyan development is the lack of correspondence of the processes of gastrulation and embryo formation with those of other vertebrates with meroblastic eggs, in particular, teleosts and reptiles and birds. Suffice to say that the posterior rim of the chondrichthyan blastoderm thickens to form a medial embryonic shield with two lateral crescent-like arms. The arms eventually close to form the embryonic axis. The embryo-forming region overhangs the yolk mass. Initially, the anterior end of the embryonic axis is a fixed point and lengthening of the axis is accomplished by posterior growth brought about by the joining together of the right and left arms of the shield Although considerable effort has been invested in the descriptive and experimental study of the problem, according to Ballard et al. (1993,

p. 328), “No consensus has been reached as to how the morphogenetic cell movements are taking place in this or any other elasmobranch fish.”

Another area of inquiry in which the advances in embryological microtechnique provided spectacular insights was the study of oogenesis. Although lampbrush chromosomes were first seen by Flemming in 1878, it was Ruckert (1892a) who first gave evidence that these structures are chromosomes. He investigated the fate of the chromosomal material during the growth of the oocyte nucleus in the ovaries of three elasmobranchs, Scyllium (=Scyliorhinus), Torpedo, and Pristiurus (=Galeus). Ruckert made three major contributions.

He was the first to devise a method of isolating intact germinal vesicles and staining lampbrush chromosomes in situ. Secondly, he described the characteristic structure of the lampbrush chromosomes, especially the lateral loops, and was the first to use the term “lampbrush.” Finally, by examining stages in oogenesis, he described the genesis of lampbrush chromosomes during early stages of oocyte growth (18 wm to 2 mm), their unique structural configuration at full development (2-3 mm), and the retraction of the lateral loops and contraction of the chromosome axis as these chromosomes give rise to normal condensed meiotic bivalents during later stages (3-13 mm) of oogenesis. Rickert’s findings were fully confirmed by Marechal (1907) in an elegantly illustrated study of two sharks and two teleosts.

Within the context of the reproductive system, some other notable contributions should be mentioned. Woods (1902) carried out a quantitative study in which he discovered that the germ cells of the spiny dogfish shark Squalus acanthias originate in endoderm associated with the yolk sac and migrate apparently by amoeboid movement to the germinal ridge of the developing gonad. Borcea’s (1905) classical study of the urogenital system contains the definitive statement on the structure of the nidamental (=shell) gland of the oviduct and its role in the secretion and morphogenesis of the egg case.

By the end of the nineteenth century, the descriptive aspects of chondrichthyan development were well known. The state of knowledge compared favorably with what was available for the more intensively studied domestic fowl and selected amphibians. Building on this foundation, future investigators would be able to formulate a series of questions based upon a functional and analytical approach to the study of development.

Comparative Embryology

Comparative studies of fish development constitute one of the glories of nineteenth century embryology. The comparative method as a formal concept in morphology has its roots in German Naturphilosophie and was extended to embryology by Dollinger and von Baer. Subsequently, Johannes Muller and others incorporated it into physiology. Already well established, comparative embryology was thrust into a position of prominence with the publication of Darwin's On the Origin of Species. Darwin stated that much could be learned about evolution through embryological studies. Fritz Mller (1864) and Haeckel (1866) considered that embryology, especially comparative embryology, was paramount in understanding evolution, i.e., "ontogeny recapitulates phylogeny". The new evolutionary impetus combined with the pre-existing research agenda drove comparative studies of fish development ahead at a rapid clip. Fishes provided a intellectual feast for evolutionary comparative embryologists. The four major groups of fishes occupy key positions in vertebrate evolution. Agnaths (hagfish, lampreys) are the most primitive representatives of the craniate-vertebrate lineage. Chondrichthyans (sharks, rays, chimaeras) are among the oldest extant jawed vertebrates and incorporate the definitive vertebrate Bauplan and developmental program. Sarcopterygians comprise lungfishes, coelacanths, and tetrapods. The Actinopterygian or ray-finned fishes (Polypterus, sturgeons, gars, Amia, and teleosts) represent a group of bony fishes that evolved in a lineage that culminated in the teleosts, one of the most successful groups of vertebrates. Not only was there evolution at the phyletic level, but the process of development itself underwent evolutionary change. Once embarked upon the comparative venture, the availability of study material became a problem. Many fishes of interest were found only in the fresh waters of the tropics, and some, such as the Australian lungfish Neoceratodus, were only discovered in the last quarter of the nineteenth century. Coincident with the rise of interest in comparative embryology, geographical regions that harboured fish species of interest became more accessible, the fishes themselves became better known, and the technology of travel improved. Even so, studying the embryology of fishes under primitive field conditions was a major task. Individuals such as Budgett, Kerr, and Semon are part of the golden age of exploration.

The hagfishes and lampreys are the most primitive representatives of the craniatevertebrate lineage. Both are jawless. Hagfishes are common, marine, bottom-dwelling fish in cold, inshore waters and in deeper waters off continental shelves. Recently, hagfishes have been considered to be the sister group to all remaining vertebrates. Virtually everything that is known about their embryonic development is derived from collections of Eptatretus (=Bdellostoma) stout eggs obtained by Bashford Dean and Franz Doflein in Monterey Bay, California, on the Pacific coast of North America in the 1890's. For all intents and purposes, there have been no modern embryological studies of hagfish except for Gorbman (1983), Gorbman and Tamarin (1986), and Wicht and Northcutt (1992), who used Dean's material. For this discussion, research on the Atlantic hagfish Myxine will not be considered because there is so little of it. Doflein (1898) and Dean (1898, 1899) described embryonic development. Dean's 1899 extraordinary monograph gives a definitive account, sumptuously illustrated with colour lithographs, of hagfish development from early cleavage through hatching. Unfortunately, because the monograph is not readily accessible, this research has been overlooked. Most of Dean's account is based on the observation of whole mounts. Sections of blastula and gastrula stages are illustrated.

The eggs are large ellipsoids about 22 mm in length by 8 mm in width. They are heavily yolked, and there is a germinal disc at the animal pole. Cleavage is meroblastic. The process of embryo formation coincides with the epibolic movement of the blastoderm over the yolk mass. The equivalent of a germ ring is present and the embryo is formed along an anterior-posterior axis as the blastoderm moves over the yolk mass. The only section of a gastrula reveals the absence of bottle cells, although this could be a function of the plane of section. Both a cellular "periblast" and an underlying yolk syncytium are depicted. The neural tube is formed by the closure of the medullary folds. Much information is given on organogenesis and the later phases of embryonic development. Additional descriptions of early cleavage stages are given by Gudger and Smith (1931). A description of development of the head, brain and associated organs based on the analysis of sectioned material was published by von Kupffer (1899).

Lampreys are widely distributed throughout the waters of the northern and southern temperate zones. They were known to zoologists such as Home, Rathke, and Miller who worked on aspects of their reproductive anatomy. It is surprising, then, that embryological research was only undertaken in the mid-nineteenth century and that comprehensive studies only appeared in the latter part of the nineteenth and in the early twentieth centuries. The first substantive description of lamprey development was that of Schultze (1856), who amongst other observations demonstrated the somewhat unequal holoblastic cleavage of the egg. Aspects of gastrulation were described by Balfour (1880), Goette (1890), and Hatta (1907). Both Goette and Hatta demonstrated the presence of bottle cells and involution of surface cells through the blastopore. Other contributions of note are those of Scott (1887), von Kupffer (1890), and Selys-Longchamps (1910). Older literature on lamprey development is listed in Ziegler (1902). Piavis (1971) provides a modern account of development and a list of the more recent literature.

Two main lines of evolution are recognized within the class Chondrichthyes: the holocephalans and elasmobranchs. When chondrichthyan development was discussed previously, only the elasmobranchs (sharks, rays, and skates) were considered. Now the holocephalans (chimaeras or ratfishes) will be considered. Chimaeras are oviparous, bottom-dwelling marine fishes usually found in the deep waters off the continental shelf, except in those regions where the inshore waters are cold, e.g., the northwest coast of North America and New Zealand. Because of their deep water habitat, it has been difficult to procure their eggs for study. The major sources of information on chimaeroid development are Schauinsland's (1903) and Dean's (1906) monographs. Schauinsland studied Callorhinchus mil, whereas Dean's study was mostly based on the eggs and embryos of Hydrolagus (=Chimaera) colliei collected in Monterey Bay, California. Dean's monograph is remarkably complete. Dean begins with an account of the breeding habitats of Hydrolagus and egg case formation and then describes in detail development from fertilization through hatching. Chimaeras are oviparous. Their eggs, moderately large, oblate spheroids (25 x 20 mm), are enclosed in an elaborate egg case composed of structural proteins that differ considerably from those of elasmobranchs. Cleavage and subsequent development is meroblastic and closely resembles that of sharks. A closely spaced series of stages from early cleavage through the mid-phases of development is illustrated with surface views of the blastoderm and drawings of the embryos. This series is supplemented with many illustrations of sectioned material, especially from the blastula and gastrula stages. One developmental characteristic that sets chimaeras apart from elasmobranchs is the pronounced elongation of the post-anal region of the body, which begins at the tailbud stage and soon gives the young embryo an almost eel-like appearance. The studies of Schauinsland and Dean remained unchallenged for many years. Only in the late twentieth century has interest in chimaeroid development revived.

The osteichthyans or bony fishes include the sarcopterygians and actinopterygians. The sarcopterygians (fleshy-finned fishes) are now considered to include the tetrapods, two extinct groups of fishes, and two extant groups of fishes: coelacanths and dipnoans (lungfishes). These fishes are of considerable interest because the "lungfishes, coelacanths, or both together are the closest living relatives of tetrapods".

The living coelacanth Latimeria chalumnae is the sole surviving representative of a once common group. The only known population occurs in the Comoros Islands off the east coast of Africa. No embryological research was carried out on Latimeria during the nineteenth century because it was not discovered until 1938. However, in 1975 Smith et al. discovered that Latimeria is viviparous when they found five nearterm pups in the oviduct. Up to this time only two gravid females have been collected. Wourms et al. (1991) described aspects of viviparity and the maternal-embryonic relationship using the embryos and maternal tissues from the Smith et al. (1975) specimen. This material was also used by Bemis and Northcutt (1991) in their study of the innervation of cranial muscles and Northcutt and Bemis (1993) in their study of cranial nerves. Fricke and Frahm (1992) described full-term pups obtained from the second gravid female. Coelacanth eggs are enormous, up to 9 cm in diameter and with a dry weight of 184 g. They are heavily yolked and contain yolk platelets. Although no information is available on early embryonic development, it is reasonable to assume that development is meroblastic. Because the coelacanth is on the verge of extinction, it is highly unlikely that any substantial body of embryological knowledge will be forthcoming.

The dipnoans or lungfishes comprise the other extant group of sarcopterygian fishes. There are three genera of living lungfishes: Protopterus in Africa, Lepidosiren in South America, and Neoceratodus in Australia. Lepidosiren and Protopterus were discovered nearly simultaneously in the late 1830's and Neoceratodus in 1870. Lungfishes as "intermediate forms" readily fitted into the emerging evolutionary paradigm of the nineteenth century. Thus, the study of lungfish development became one of the most sought after goals of comparative embryology. It entailed expeditions to three continents that involved considerable risk and in some instances loss of life, viz. Budgett.

The major accounts of lungfish development appeared during a brief sixteen-year span at the end of the nineteenth and in the early twentieth centuries. The embryology of the lungfish Neoceratodus was first described by Semon (1893) in a series of 47 stages that extend from fertilization through post-hatching juveniles. This report, which was based on the description of external features, was complemented by a study of gastrulation based on the analysis of sectioned material. Much of the previous research was summarized and new material was incorporated into a publication that was part of Keibel's series of normal tables of development and others have re-investigated the development of Neoceratodus. In much the same fashion, Kerr (1900) published the first account of the development of the South American lungfish, Lepidosiren. An account of external development of closely spaced states from fertilization through post-hatching juveniles was given. Unlike the Australian lungfish, embryos of Lepidosiren develop external gill filaments before hatching and retain them for at least 40 days. A second part of Kerr's account included a comparison with the development of the African lungfish Protopterus. The first account of its development was given by Budgett (1901). This preliminary report contained information on the reproduction and development of other African fishes, most notably Polypterus. Budgett died in 1904 from malaria contracted in Africa. Kerr, Assheton, and others acted as Budgett's scientific executors, ensuring the publication of his research. The definitive report on the development of Protopterus was made by Kerr in the same volume of the Keibel series that contained the normal table of Lepidosiren development (Kerr, 1909). Thus far, little has been said about lungfish development. Eggs are approximately 3 mm in diameter with a moderate amount of yolk. Cleavage is holoblastic and unequal. A blastopore develops in a subequatorial position. Bottle cells are present. Surface cells involute through the blastopore. The neurula is characterized by a flat neural plate that folds to form the neural tube. In general, lungfish development, especially that of the Australian lungfish, bears an uncanny resemblance to that of urodele amphibians. Caution, however, is advised because of the need for a fine-grained analysis of the early (blastula-neurula) phases of development using contemporary methods. The actinopterygian or ray-finned fishes are of particular interest to comparative embryologists. They constitute a well-defined evolutionary lineage that culminated in the teleosts, one of the most successful, diverse, and species-rich groups of vertebrates. Based on their evolutionary relationships, they provide a unique opportunity for investigating the transition from the primitive, holoblastic pattern of fish development to the specialized meroblastic pattern characteristic of teleosts. The actinopterygian fishes comprise three subclasses: Cladistia, Chondrostei, and Neopteryg.

The Cladistia are the most basal group of the surviving cladistians are represented by two genera, Polypterus and Calamoichthys, which are in the order Polypteriformes. In some classifications, e.g., Nelson (1994), the Polypteriformes are placed in the subclass Chondrostei together with the sturgeons. Both Polypterus, the sturgeons and paddlefish are considered to be part of the lineage that led to the teleosts. Polypterus, the bichir, and Calamoichthys are from Africa. Little or nothing is known about the development of Calamoichthys.

The situation is somewhat better with respect to Polypterus. The first, very fragmentary report of its development was made by Budgett (1901). Following Budgett's death in 1904, Kerr worked up Budgett's collections and notes and published the only extensive account of the development of Polypterus. Almost all aspects of development from fertilization through hatching and also post-hatching larvae are described in 36 stages. Eggs are small, slightly flattened spheres (0.9 x 1.3 mm along the major and minor axes), moderately yolked, and contain yolk platelets. Cleavage is holoblastic and unequal. Unfortunately, information on the critical events of gastrulation is scanty. At the onset of gastrulation, what appears to be a very wide blastopore occupies an equatorial position. Subsequently, the lateral margins of the blastopore extend and meet to form a circle nearly as large as the circumference of the egg. This marks the advancing boundary of the surface cells of the animal pole during epiboly. Epiboly is pronounced and initially a very large yolk plug is formed. Subsequently, it is reduced in size.

The marginal quality of the single section of a gastrula stage illustrated by Kerr (1907) does not allow comment on the presence or absence of bottle cells. Better sections through early neural plate stages confirm the presence of a blastopore that is continuous with a large archenteron. During neurulation, a flat neural plate folds to form a neural tube. The embryo passes through typical tail bud stages and eventually hatches as a larva with external gill filaments. According to Arnoult (1964), the rate of development is quite rapid, i.e., 48 hours from fertilization to hatching. Although there are differences, the development of Polypterus resembles that of the sturgeon. Obviously, Polypterus is a prime candidate for further study. The prospects are excellent for it is in the aquarium trade and can be spawned in captivity. In point of fact, Bartsch and Britz (1996) have just published a popular account of spawning and development that includes scanning electron micrographs of the neural plate stages. Some of their material was used in Piotrowski and Northcutt's (1996) study of cranial nerves.

The subclass Chondrostei has one extant order, the Acipenseriformes, that contains the sturgeons and paddlefishes. They are considered an offshoot of the lineage that led to the teleosts. Sturgeons occur in North America and Eruasia. Paddlefish are restricted to the United States and China. The development of sturgeons was first outlined by Kowalewsky et al. (1870) for the European sterlet (Acipenser ruthenus). A more detailed account of early development and embryogenesis was given by Salensky (1878, 1881). Somewhat later, similar studies were carried out on the eggs of the North American Atlantic sturgeon A. oxyrhinchus.

Recently, Ginsburg and Dettlaff (1991) and Dettlaff et aL (1993) have presented a detailed contemporary account of the development of the Russian sturgeon A. gueldenstaedt from fertilization through hatching. Eggs are 3.0-3.5 mm in diameter. Cleavage is holoblastic. A dorsal lip develops in an equatorial position as the apices of bottle cells contract. Involution of surface cells takes place through the blastopore. The neurula stage is characterized by a flat neural plate that subsequently folds to form the neural tube. According to Bolker (1993a), gastrulation is similar to that of the amphibian Xenopus. Morphogenetic movements have been analysed by Ballard and Ginsburg (1980). The mechanisms of gastrulation and mesoderm morphogenesis have been investigated by Bolker (1993a, 1993b). The embryology of the paddlefish (Polydon) did not become known until the 1960s. The early-and mid-phases of paddlefish development closely resemble that of sturgeons. During the later phases of development, embryos develop their characteristic elongated rostrum. A scanning electron microscopy study by Bemis and Grande (1992) has substantially extended the description of Polyodon development.

The subclass Neopteryg contains three divisions, viz. Ginglymodi (gars), Halecomorphi (Amia), and Teleostei. Gars (Lepisosteus) occur in eastern North America and Central America. The first account of the development of gars was by Alexander Agassiz (1878), a report that dealt exclusively with late embryonic and post-hatching stages. Using material provided by Agassiz, Balfour and Parker (1882) gave an account, albeit fragmentary, of some of the earlier phases of development. Most of their report, however, contains the first detailed account of embryogenesis. The egg is about 3 mm in diameter and heavily yolked. Cleavage is meroblastic. Sections of embryos reveal that the spinal cord originally is a solid rod that secondarily becomes hollow. Dean (1895) provided critical information on the early development of Lepisosteus from fertilization through gastrulation and compared its development with that of the sturgeon. Dean was the first to provide a timed series of developmental stages for both fishes. In addition to providing surface views of the external features of early development, Dean also based his descriptions on sections of a developmental series. Eycleshymer (1899) provided additional information on the cleavage process.

Recently, Long and Wourms (1991) examined early development with scanning electron microscopy. From these studies, the following view of early development of Lepisosteus emerges. Meroblastic cleavage produces a blastoderm atop a large, yolk sphere in which abortive cleavage furrows are present. Further studies are needed to ascertain whether the yolk mass undergoes any division. Epiboly carries the blastoderm margin toward the vegetal pole preceded by microvilli-bearing zone of the yolk surface similar to the yolk syncytial layer of teleosts. As epiboly of the blastoderm reaches the equator, a blastopore begins to form at one point along the rim of the blastoderm. Bottle cells are associated with the dorsal lip of the blastopore. By this time, cells of blastoderm are organized into an embryonic shield and a wide germ ring. As epiboly proceeds to completion, the germ ring narrows, the embryonic shield elongates, and a shallow, neural groove forms similar to that of teleosts. Amia calva, the bowfin, is the sole extant representative of the order Amformes. It is confined to the freshwaters of eastern North America. Its development was made known in a flurry of papers that appeared during 1896-1906. Dean (1896a) reported on early development, i.e., fertilization through gastrulation. External views of successive stages were supplemented with sections of the corresponding stages. The later stages of embryonic development as well as the development of the post-hatching larvae were reported on separately. Almost simultaneously, Whitman and Eycleshymer (1897) reported on reproduction and the cleavage stages of development. Again, external views of the whole egg were supplemented with sections of the corresponding stages.

It is interesting to note that Whitman raised the issues of both priority and unfair rivalry with Dean because Dean obtained eggs of Amia from the same small lake in Wisconsin in which Whitman had previously discovered spawning colonies. Ballard (1986a,b) has published a series of stages and rates of normal development and has investigated morphogenetic movements. He has also produced a provisional fate of development in Amia. Superficially, the development of Amia resembles that of Lepisosteus but with several important differences. The egg is an elongated sphere (2.2 X 2.8 mm along the major and minor axes). A germinal disc occupies about one-third of the egg volume. Cleavage is holoblastic but quite unusual. The yolk mass divides slowly and with apparent difficulty whereas the germinal disc is rapidly divided into many small cells. In late cleavage, a blastoderm surrounds a yolk mass divided into giant blastomeres. Cells of the blastoderm can be distinguished as enveloping layer cells, epiblast, or hypoblast.

As epiboly commences, morphogenetic movements within the blastoderm lead to the formation of the germ ring and embryonic shield. The early phases of embryo formation take place during epiboly of the blastoderm. One of the most significant differences between the development of Amia and Lepisosteus is that Amia lacks both a blastopore and bottle cells. It is disconcerting that Lanzi (1909) illustrated sections of a groove at the leading edge of the blastoderm that contains cells that appear to be bottle cells. For obvious reasons, ultrastructural studies are in order. Collazo et al. (1994) discuss in detail the phylogenetic implications of these comparative embryological studies of primitive actinopterygian fishes with respect to teleost gastrulation.

Teleosts constitute the remaining major group of actinopterygian fishes. The history of teleostean embryology has been discussed. The main emphasis of early studies was placed on defining the meroblastic pattern of teleostean development and piscine features of embryogenesis and organ and tissue formation. Some attention was given to the diverse embryonic adaptations for viviparity in different taxa. By the end of the nineteenth century, it was apparent that the early development of most teleosts was very similar but that mid-and late-phases diverged, yielding diverse embryonic morphology and larval forms adapted to specific life styles and habitats. Thus, the later phases of embryonic development, larval development, and life histories became the subjects of comparative studies of teleostean development during the late nineteenth century and the twentieth century.

Several factors favoured the growth of a comparative approach, especially the advent and rise of marine biology, biological oceanography, and fisheries biology. Fishes are the dominant group of marine vertebrates and are commercially important. Thus, the study of life histories and the embryonic and larval stages of inshore, pelagic, and deep-sea fishes became a research agenda for the marine sciences. The results were often quite spectacular, e.g., the leptocephalus larvae of eels and the bizarre larvae of the stomiatoid fish Idiacanthus with its stalked eyes and trailing gut.

Studies of this type were pioneered by Alexander Agassiz (1882) and Ryder (1884a) in the United States, McIntosh and Prince (1890) in Great Britain, and Ehrenbaum (1909) in Germany, amongst others. Similar activities were undertaken at limnological stations and in freshwater fisheries biology. Popularization of the aquarium hobby and the successful reproduction in captivity of freshwater fishes imported from Africa, South America, and Asia such as the zebrafish, Danio rerio, established another source of material for comparative studies. Given their numbers and diversity, a wealth of information, mostly of a descriptive nature, was produced on the development of a great many species of teleosts.

Increased activity in the study of teleostean development was accompanied by the elaboration of several concepts and theoretical approaches which both aided and impeded further progress. In 1877, Haeckel set forth his Gastrea theory. Based on his observations of teleostean development, he applied the concept to both teleosts and chondrichthyans. Much of the literature on fish development from his time to the present has been concerned with the morphogenetic movements and embryo-forming mechanisms that are associated with gastrulation, a literature too voluminous to review. One major problem that arose from overzealous and inept efforts to use embryological studies to support evolutionary theory was the attempt to homologize the events and processes of gastrulation in teleosts and chondrichthyans with amphibians and amniotes. In this respect, it was Goette (1873) who first proposed that involution, i.e., flowing of a sheet of cells over an inpocketing such as a blastopore, has a prominent role in teleost gastrulation.

At the risk of oversimplification, many of these studies were comparative in name only. There was little or no conceptual basis for interpreting the mass of information. Attempts, such as that of Schultz (1910), to establish a well-defined conceptual basis for comparative embryological research were overlooked. (In passing, Schultz's 1910 monograph appears to be a direct precursor of de Beer's 1930 study of embryology and evolution.) Although the body of evolutionary theory was available, the influence of Haeckel's biogenetic law interfered with the productive application of it to many developmental issues.

A growing anti-evolutionary bias tended to discourage this type of inquiry. Studies of individual species were often exceedingly well done, and the development of a single organ system was often compared in several different taxa and interpreted in the context of the evolutionary paradigm. Examples of the latter include Ryder's (1886c) study of heterocercy and the evolution of the fins and fin rays of fishes and Eigenmann's (1903) study of the abortive development of the eye in blind cave fishes. Eigenmann compared the development and degeneration of the eye in the blind cave fish Amblyopsis with that of the surfperch Cymatograster, a fish with a well-developed eye. In Amblyopsis, development of the eye begins normally, but then the lens degenerates and the entire eye regresses. Recently, Jeffery (1996), using the embryos of another blind cave fish Astyanax, has shown that programmed cell death of the lens eliminates lens induction from the cascade of inductive interactions required for normal eye development.

In general, however, the comparative method was not used to explore the evolution of development and the factors that brought about evolutionary change. It was not until the maturation of evolutionary thought in the twentieth century that a new set of concepts and approaches were produced and used to exploit the potential of comparative studies. Paramount among these are: (1) revived interest in the evolution of development and the role of development in evolution; (2) application of evolutionary theory to life histories; and (3) attempts to order the bewildering diversity of fish reproduction in terms of systematics or reproductive guilds.

Analysis of Development

The term "analysis of development" comprises a set of conceptual and methodological approaches that are used to study development, for example, functional morphology, biomechanics, physiology, biochemistry, and cellular biology. These approaches possess a strong experimental and causal bias. It is difficult to say when such analytical approaches to the study of development were introduced.

Although its origins are obscure, the experimental-analytical approach certainly was recognizable in the eighteenth century studies of Spallanzani (1729-1799) on insemination, Trembley (1710-1784) on hydra regeneration, and Broussonet (1789) on regeneration in fishes. By the nineteenth century, experimental physiology under the aegis of Claude Bernard and Johannes Muller had become a formalized science. Muller was a pivotal figure for he was both a physiologist and an embryologist. It is probably through him and his many students that an analytical-experimental approach was transferred from physiology to development and became established in the mid-nineteenth century. For example, Haeckel (1869), a student of Mller, divided the larvae of the siphonophore Crystallodes in two and obtained smaller, but perfect larvae. It was only later in the nineteenth century that a more narrowly defined, conceptual approach emerged in the Entwicklungsmechanik of Roux and the mechanistic views of Loeb. Here I will discuss a few examples that demonstrate the analytical approach to fish development.

The introduction of new embryological techniques opened up new areas of inquiry especially at the cellular level, and led to a series of important observations, a few of which will be discussed here. The organization of the teleostean egg, which differs considerably from that of most other vertebrates, offers unique opportunities for the study of fertilization. The mature ovum is invested with an extracellular matrix, the egg envelope. At the animal pole of the egg, there is a funnel-shaped channel through the egg envelope termed the micropyle. Fertilization is accomplished by the passage of a single spermatozoan through the micropyle and its subsequent fusion with the ooplasm of the germinal disc. The micropyle was first described by von Baer (1835) and reinvestigated by Doyere in 1849, who appreciated its significance in fertilization but did not have any experimental evidence for its role.

Ransom (1867), in an extraordinary paper on the eggs of teleostean fishes, not only experimentally demonstrated fertilization, but also conducted experimental and analytical studies of egg organization and post-fertilization ooplasmic contractility. In a series of experiments, in which he was able to restrict or facilitate access of spermatozoa to the micropyle, Ransom (1867) observed the passage of a single spermatozoan through the micropyle and the subsequent events of the fertilization reaction, namely, changes in structure of the micropyle, disappearance (discharge) of the cortical "granules," the appearance of the perivitelline space, and the movement of cytoplasm into the germinal disc. He did not document the cytological details of spermegg fusion. This was done later by Hoffmann (1881) who depicted aster formation, pronuclear fusion, and mitosis. Ransom (1866, 1867) also carried out experimental studies to ascertain the physical and chemical properties and structural organization of maturing oocytes, mature ovum, and fertilized egg. He consciously approached the egg from a cellular point of view. The eggs of some fishes, especially the stickleback Gasterosteus and the pike Esox, exhibit strong rhythmic contractions of the yolk mass shortly after fertilization and throughout the early cleavage stages. Contractile waves, originating at the animal pole, bring about transitory deformations of the yolk mass.

Ransom (1867) found that the contractions are independent of fertilization and appear to reside in what is now known as the yolk cytoplasmic layer. He conducted elaborate experiments to ascertain the "conditions" that regulate both contraction of the yolk mass and cytokinesis. The effects of a number of factors, viz. metabolic poisons, temperature, electricity, carbon dioxide, and oxygen were determined. He devised a system whereby he could subject eggs to electrical stimulation while simultaneously observing them through a microscope. Electrical stimulation excited peristaltic waves that were distinguishable from normal ones by their site of origin, rapidity of formation and progress, and greater deformation. Cleavage was not affected. Cold temperatures delayed the fertilization reaction and slowed cleavage whilst moderately warm temperatures accelerated cleavage. High temperatures irreversibly arrested contractions and cytokinesis. Cyanide had no effect on yolk contraction, but delayed cytokinesis. Morphine and chloroform inhibited contraction.

Carbon dioxide was found to be a potent inhibitor of both contraction and cytokinesis. On the basis of a series of experiments on oxygen requirements, he concluded: (1) oxygen at low levels is necessary for contraction and cytokinesis; (2) cytokinesis requires more oxygen than do yolk contractions; (3) embryonic stages involving cell multiplication and differentiation require much more oxygen than do early cleavage stages; and (4) the embryonic heart has a higher tolerance to oxygen deficiency than does striated trunk muscle.

In a separate publication that also dealt with rhythmic yolk contractions, Ransom (1866) reported on the isolation of a protein "allied to myosin" in the yolk of all fishes, amphibia, and birds that he had examined. He did not, however, state that myosin is involved in yolk contraction. This apparent omission may be an instance of understatement.

In addition to his fertilization and contractility research, Ransom also described and isolated the germinal vesicle, the enlarged nucleus of early oocytes, and illustrated the multiple nucleoli within it. His pioneering efforts were not fully realized until the twentieth century. Most of the research during the latter half of the century involved histological studies of oocytes and eggs. In this respect, there were some interesting developments in the study of nuclear-cytoplasmic relationships and chromosomes. In a histological study of oogenesis, Van Bambeke (1893) depicts lampbrush chromosomes without recognizing them and reports on the "elimination of nuclear elements" from the nucleus into the ooplasm of early oocytes. It appears that he was describing ribosome production and storage. Lampbrush chromosomes were first described in the eggs of elasmobranchs (vide infra) and subsequently discovered in teleosts. The discovery of the behaviour of chromosomes during cell division was soon extended to include the cleavage stages of teleosts.

The ability to maintain and manipulate cells in vitro is an important tool of modern biology. Ross Harrison (1907) is usually credited with the "discovery" of cell-tissue culture. However, he had several forerunners, one of whom was Wilhelm His. His (1899) isolated single blastomeres and clumps of blastomeres from salmonid blastulae and maintained them in the transparent fluid portion of the egg yolk. Under these conditions, he was able to make longterm observations on the in vitro behaviour of blastomeres. He compared the behaviour of blastomeres from 0-4 day post-fertilization eggs (cleavage stages) with blastomeres from 5-12 day post-fertilization eggs (gastrula stages). He observed "amoeboid" movement and the extension and retraction of finger-like processes which contained clear cytoplasm (=hyaloplasm) in the 5-12 day blastomeres, but not in the 0-4 day blastomeres.

In the later stage blastomeres, he also observed a transition in organization from single cells with finger-like processes, loosely arrayed in clusters, to round cells that are joined together to form epithelial sheets. He compared amoeboid movement and contractility of blastomeres to muscle contraction. The cytoplasmic organization of living and fixed blastomeres was shown to be similar. The leading edge (lobopodia) of the blastomere consists of hyaline cytoplasm which may contain elements of a sparse fibrous reticulum (ER, cytoskeleton?) whereas the perinuclear cytoplasm is dense and contains both granular elements and a fibrous reticulum. Dynamic changes in the regional organization of the cytoplasm in living cells was noted. These observations were illustrated with both drawings and photomicrographs of sectioned cells. The in vitro observations of His are remarkably similar to those carried out by Trinkaus on the blastomeres of Fundulus many years later. The first attempt to analyse the mechanism of embryo formation in teleosts was undertaken by Lereboullet in a set of descriptive and experimental studies of teratology. He used different temperature regimes and chemicals to induce abnormal development, and also tested the effect of pressure on development. Although he used over 200,000 eggs in his experiments, the results were inconclusive because the incidence of induced abnormalities in treated eggs was no greater than that in untreated eggs.

Oppenheimer (1936) speculated that Lereboullet worked with too many eggs and should have monitored each egg. Careful individual analysis of the many "spoiled eggs" in the experiments would have produced more conclusive results. Lereboullet's descriptive studies were more fruitful. By tracing the progress of developmental anomalies, especially duplicated or partially duplicated embryos from their first visible origin, Lereboullet obtained insight into both their development and the normal process of embryogenesis. These studies caused him to focus on the embryo-forming dynamics of the germ ring. Eventually, he concluded that embryo formation in teleosts took place by the process that His (1874) later called concrescence.

The pioneering efforts of Ransom and Lereboullet were capitalized on by Loeb, Stockard, and others. They used a variety of physical and chemical parameters either to determine the physiological conditions required for normal development or to investigate the mechanisms of developmental regulation by perturbing those conditions. Loeb began his work on fish embryos in the 1890's primarily using the cyprinodontid Fundulus, found in abundance at Woods Hole. In a series of early experiments, he found that embryos, the hearts of which had been immobilized by potassium chloride, developed structurally normal circulatory systems but ones lacking blood circulation. Chromatophores that normally migrated onto the surface of the vitelline vessels of the yolk sac failed to do so. Loeb suggested that migration would only occur if there was blood circulating in the vessels. Later, used his research on the effects of ions on embryos of Fundulus as the basis for his theory of physiologically balanced salt solutions. Stockard (1906, 1907) studied the effects of magnesium and lithium ions on the development of Fundulus in order to compare them with the development-perturbing effects already well known in sea urchins and frogs. At high ionic concentrations, cyclopean monsters with a single median eye were produced. By varying ion concentrations, stages intermediate between normal and cyclopean could be produced. The ionic perturbation interfered with the separation of the early, single eye-forming field into the right and left eye anlagen.

Although radioactivity was only discovered in the latter years of the nineteenth century, radioisotopes were soon employed in embryological research. For example, Tur (1896) studied the deleterious effects of radium radiation on catshark Scyliorhinus development, and Oppermann (1913) described a variety of developmental abnormalities that resulted from the fertilization of trout eggs with radium-irradiated spermatozoa.

Fundulus has been a favourite experimental animal at Woods Hole and other marine laboratories on the east coast of the United States for over 100 years. Given its popularity, it is natural to inquire into its history. The first accounts of the development of Fundulus were given by Alexander Agassiz (1882) and Ryder (1886b). Boyer (1892) was the first to specifically state that he had obtained eggs by stripping ripe females and artificially fertilizing them, and he also gave a more extensive account of the early stages of development. Who, then, introduced the eggs and embryos of Fundulus as experimental material and developed the methods for obtaining them? I speculate that John Ryder was the driving force behind Fundulus.

Ryder was an embryologist's embryologist. In his brief life, he published numerous (215) papers, many of which dealt with fishes. From 1880 to 1886, he was in charge of embryological research at the U.S. Fish Commission Laboratory at Woods Hole. Much of his research on fish embryology required artificial fertilization of eggs. The Fish Commission Laboratory was far better equipped for research than was the fledgling Marine Biological Laboratory. Consequently, many embryologists such as Boyer, Morgan, and others started their work there and continued as long as the cultural atmosphere was conducive to basic scientific research. Ryder left the Fish Commission Laboratory for a professorship at the University of Pennsylvania which he held until his death in 1895. Both Morgan and Loeb were on the faculty of Bryn Mawr and in close proximity to Ryder. Did their paths cross?

In terms of what has come to be regarded as classical experimental embryology, the first experiments of consequence on embryonic fishes are based on the concrescence theory. A different approach to the study of fish embryology had been taken by His, who was more concerned with the events and mechanisms of development than their possible evolutionary significance. In his quest for a mechanistic theory, His (1874) proposed his "Konkreszenz Theorie" to explain the observed relationship between embryo formation, and the cellular material in the lateral arms of the germ ring.

As originally conceived by His, concrescence was a process by which the right and left lateral margins of the germ ring became apposed to one another in the forming embryo, so that as the two lateral margins progressively fused in a medial plane, the embryo grew backwards. In this theory, the embryo was formed by the union of the two halves of the germ ring so that successive regions along the anterior-posterior axis of the embryo were originally located in successively distal regions along the circumference of the germ ring. His applied his theory both to teleostean and chondrichthyan development. Although not implicitly stated, His's theory is a form of mosaic or determinate development. Either because it was challenging or because it was so succinctly stated, the concrescence theory soon was put to the test. The use of experiments to test a theory of development was a novel idea at that time and represents an early phase in the paradigm shift to experimental embryology.

The first person to test the concrescence theory was Kastschenko (1888), who operated on the embryos of the catshark Scyliorhinus. He carried out three sets of experiments but did not illustrate his results. In the first, he cut the germ ring on one side of a Stage 7, lancet-shaped embryo. A normal embryo developed. In another experiment, he destroyed the posterior end of the blastoderm but not the germ ring. The anterior half of the embryo developed normally, but the posterior half did not. If His's theory were correct, the posterior half of the embryo should have developed because the tissue for its formation was intact in the germ ring. In the third experiment, the entire embryo at Stage 7 was separated at its two sides from the blastoderm, severing its connection to the germ ring. A normal embryo, with both sides of the body intact, continued to develop to the three somite or tailbud stage. According to His, a defective, partial embryo should have been produced because the cellular material needed to complete embryo formation remained in the germ ring.

Kastschenko's pioneer research was soon followed by more extensive studies by Morgan (1895) and Kopsch (1896, 1898), who tested the concrescence theory in both chondrichthyans and teleosts. Kopsch's results differed somewhat from those of Kastschenko. Morgan's research is interesting not only because of its bearing on the concrescence theory, but also because of his experiments on early blastomeres. Morgan (1895) carried out two sets of experiments on the eggs of Fundulus. In the first set, he completely removed one of the first two blastomeres. The remaining blastomere divided, and a somewhat smaller blastoderm was produced.

From it a single, normal embryo formed, albeit somewhat smaller than usual. In the second set, Morgan cut the germ ring on one side of an early embryo. The embryo continued to develop normally and underwent posterior elongation even though the germ ring remained intact only on the unoperated side. The results contradict His's theory. In retrospect, His's concrescence theory was an excessively rigid statement of the process of confluence of germ-ring cells and their incorporation into the embryo. Before leaving the subject of concrescence, I wish to point out that His restated the theory in 1894, but, more importantly, in that paper he introduced physical models as a means of illustrating the mechanical basis of morphogenesis.

Other approaches to the study of developmental mechanisms were undertaken during this period. Following his pioneering application of cell lineage analysis to the development of Clepsine, Whitman attempted to take a similar approach to teleosts. Both his attempts and those of Wilson (1891) were premature and doomed to failure because there was no sure way of identifying individual blastomeres and their progeny and tracing their fate during development.

The study of the physiology of embryonic nutrition is the last topic to be considered. This aspect of piscine development had its origin in the study of viviparous fishes. In many such fishes, there is a marked difference between the small size of the egg and the large size of the full-term embryo. How is this accomplished? The problem of embryonic nutrition in viviparous fishes was recognized as early as 1624 when Schoenveld postulated that embryos of Zoarces ingested ovarian fluid as a source of nutrients. It was not until the nineteenth century that physiological and chemical studies of embryonic nutrition were attempted. Davy (1834) appears to be the first to have undertaken this line of investigation. He compared the mean weight of the egg (182 grains) of the electric ray Torpedo with that of the full-term embryo 479 grains). On the basis of the increase in weight, he concluded that the embryo received additional maternal nutrients during gestation, a correct conclusion but based on faulty data collection. Modern studies, reviewed in Wourms et al. (1988), show that there is actually a loss of organic weight during development of this fish.

The apparent discrepancy arose because wet weight instead of dry or organic weight was measured. Teleosts also were studied. In 1867, Blake analysed the ovarian fluid of a surfperch and found that proteins and lipids were the major organic constituents and hence the source of nutrition. Analysis of ovarian fluid was next carried out on Zoarces by Stuhlmann (1887) and Kolster (1905), who found it to be protein-and lipid-rich. Interest in this area of inquiry waned until it was revived by Ranzi (1932) and attracted the attention of Needham (1942).

5

Materials and Methods in Breeding

Collection of Embryos

Golden hamsters (*Mesocricetus auratus*), 10–12 wk old, were kept under conditions of constant temperature (21–23°C) and a cycle of 14L:10D (lights-on at 0500 h). They were given laboratory chow (MF; Oriental Yeast Co. Ltd., Tokyo, Japan) and water ad libitum.

The estrous cycle of each female was determined according to Orsini. In the evening of Day 4 of the estrous cycle, females were caged with males and checked for presence of spermatozoa in the vaginal smear. Successfully mated hamsters were sedated with ether and killed by cervical dislocation.

Embryos at 1-cell to blastocyst stages were flushed from the oviducts and/or uteri with Dulbecco's PBS (DPBS; Gibco, Grand Island, NY) supplemented with 0.1% polyvinyl alcohol (PVA; Sigma, St. Louis, MO; DPBS-PVA).

One-cell embryos were recovered at 1830–1930 h of Day 1, 2-cell embryos at 0530–0630 h of Day 2, 4-cell embryos at 0300–0500 h of Day 3, early 8-cell embryos at 1100–1200 h of Day 3, late 8-cell embryos at 1830–2000 h of Day 3, and morulae and blastocysts on Day 4. The zona pellucida (ZP) was removed with prewarmed, acidified DPBS-PVA (pH = 2.5) for 30–45 sec and 0.5% pronase for 1–2 min.

Decompaction and Disaggregation

Compact embryos were decompacted by incubation for 20 min in calcium-free DPBS. Embryos were disaggregated to single blastomeres by repeated pipetting with a flame-polished micropipette.

SEM Observations

Methods for preparing embryos and blastomeres for SEM observations have been reported previously. Briefly, intact embryos and dissociated blastomeres were fixed for 1 h in 3% glutaraldehyde and 0.5% paraformaldehyde in Hanks' balanced salt solution (Gibco) with 0.1% polyvinyl alcohol (HBSS). They were washed in 3 changes of HBSS and placed on small glass coverslips (6 × 6 mm) coated with 0.1% poly-L-lysine solution (Sigma). Intact embryos or disassociated blastomeres on the coverslips were postfixed in 1% osmium tetroxide in HBSS for 1 h. After being rinsed, samples were incubated in 2% tannic acid solution for 2 h, rinsed, and then reosmicated for 1 h in 1% osmium tetroxide in distilled water. Specimens were then dehydrated in a series of increasing concentrations of ethanol, critical point-dried, and sputter-coated with gold. Observations were made with a JSM5300 scanning electron microscope (JEOL, Tokyo, Japan) at an accelerating voltage at 10 kV.

Fluorescent Observations

Labelling by Fluorescein-isothiocyanate (FITC)-conjugated Concanavalin A (Con A)

FITC-conjugated Con A (FITC-Con A) was used to label the position of the microvillous pole on intact and isolated cells. Blastomeres were fixed in 4% paraformaldehyde, washed, and incubated for 15 min at room temperature in FITC-Con A (100 ìg/ml in PBS + 0.4% BSA + 0.02% sodium azide). Samples were examined using a fluorescence microscope (IMT2-RFC; Olympus, Tokyo, Japan) equipped with a DMIB filter set (Olympus) for FITC.

Labelling of Nuclei and Cytoskeleton

Preparation of the fixative was based on protocols of Albertini and Clark and Aman and Parks with minor modifications. Microtubule stabilization buffer (MTSB) was prepared as a stock solution (pH = 6.9) containing 0.5 M PIPES, 25 mM magnesium chloride, and 12.5 mM EGTA. Fixative was prepared by mixing 2 ml MTSB stock solution with 0.01% aprotinin, 1 mM dithiothreitol, 50% deuterium oxide, 1 mM paclitaxel, 5% Triton-X, and 5.4% formaldehyde, and made up to 10 ml with distilled water. Intact embryos or dissociated blastomeres were incubated in the fixative at 37°C for 1 h and then left overnight at 4°C before fluorescein staining. The fixed cells were exposed to anti-â tubulin primary antibodies (1:200; Sigma) at 4°C overnight, then washed with wash solution containing 2% BSA, 2% goat serum,

0.2% milk powder, 0.2% sodium azide, and 0.1% Triton-X in calcium-free DPBS. The cells were subsequently incubated with the FITC-conjugated secondary antibody (1:200; Sigma) at 37°C for 2 h and then stained with rhodamine-phalloidin (1:1000; Molecular Probes, Eugene, OR) for microfilaments for 1 h and Hoechst 33342 (10 mg/ml) for DNA.

Samples were viewed on an Olympus microscope (BX-FLA; Olympus). A U-MWIB filter set (Olympus) was used for FITC, a U-WIB set (Olympus) for rhodamine, and a U-MWU set (Olympus) for Hoechst 33342. A cooled CCD video system (ImagePoint; Photometrics Ltd., Tucson, AZ) was used to obtain images on the personal computer, and colour adjustment was performed by IPLab-Spectrum P software (Signal Analytics Corporation, Vienna, VA).

Data Analysis

A total of 443 individual blastomeres from 128 hamster embryos before (1- to 7-cell stage), during (8-cell stage), or after (9- to 12-cell stage) compaction were processed for SEM. Pre-compacted embryos had relatively spherical blastomeres, whereas compacted embryos contained more flat blastomeres. Decompaction in calcium-free medium, however, made observations of microvilli distribution on the surface easier. SEM observations revealed that 1- to 7-cell-stage embryos as well as the majority of the early 8-cell embryos showed a dense and uniform distribution of microvilli over the cell surface, although the distribution density of microvilli in the 8-cell embryos decreased relative to younger embryos. Ten percent of the blastomeres (n = 80) at the early 8-cell stage showed microvillous polarization. Late into the 8-cell stage, polarized blastomeres were noted more frequently (28%, n = 128). In contrast to the mouse embryo, the microvillous pole in most polar blastomeres of the hamster embryo was found in the basal region of cell-cell contact, while fewer microvilli were present at the apical, outward-facing surface of the each blastomere. Dissociation of blastomeres clearly allowed surface characteristics of the nonpolar and polar cells to be demonstrated. Mean population of polarized blastomeres increased at the late 8-cell stage (28%) and at the morula stage (35%). Some blastomeres (9–15%) at and after compaction were difficult to isolate, and their polarity could not be determined.

Labelling by FITC-Con A

Labelling patterns with FITC-Con A did not vary with the developmental stage of the embryos (data not shown). Both the intact embryos and the dissociated cells were surface-labeled with FITC-Con

A irrespective of the cell stage; however, an accumulation of FITC label was frequently noted in that region of cytoplasm adjacent to another blastomere.

These features were coincident with the observation by SEM, which revealed that microvillous poles were generally faced inward and were situated directly between adjacent blastomeres in the hamster embryo, in contrast to those of mouse 8-cell embryos.

Labelling of Nuclei and Cytoskeleton

An outward migration of blastomere nuclei was observed with the progress of embryo development at compaction. At 1- to 7-cell stages, all blastomere nuclei were located in the center or near the center of the cell or occasionally in the basal half closer to the point of cell contact. At the early 8-cell stage, most nuclei were similarly located in the center of the cell, but some had begun to migrate outward, toward the cell surface. By the late 8-cell-stage, 80% (n = 49) of blastomeres were found with their nuclei localized to the apical part of the cell. Such features were also noted in the outer blastomeres of morula-stage embryos.

Discussion

SEM of hamster 1- to 7-cell embryos and their dissociated blastomeres showed a uniform distribution of microvilli. Onset of polarization of microvillous distribution was noted at the early 8-cell stage. More polar cells were observed as compaction progressed. Con-A staining of hamster embryos regardless of stages did not reveal any polarity of their cells in the present study. Previous studies in mice reported that binding sites of ligand, such as Con A and peanut agglutinin, may indicate the presence of microvilli at the heavily labeled pole. However, we could not detect a polarity in FITC-Con A binding in these blastomeres, which displayed a ring stain irrespective of the embryo's developmental stage. Attempts to detect polarity using FITC-Con A were also unsuccessful for cattle and rabbit embryos, but limited success was achieved in cattle when laser confocal microscopy was employed. We found in the present study that an accumulation of labelling was restricted to the internal surface membranes of blastomeres in the cleaved embryo. Some authors observed heavy labelling of FITC-Con A at the cleavage furrow of the mouse embryo; probably this is the area associated with the actin filaments of the contractile ring. The reasons for the differences between our results in hamsters and previous studies in mice are not clear, although this difference can probably be attributed to the different mechanisms of embryo compaction among different species. It has been reported that microvilli include a parallel linear array of microfilaments. In the present study, microfilaments were observed at a much higher density in the region of cell-cell contact than in the outward free cell surface. However, a strong FITC-Con A stain and higher density of microfilaments in this contact region may not always indicate a higher density of microvilli polarity, because we obtained similar results from pre- and post-compacted embryos by both fluorescence staining methods, but not by SEM as described above. Reeve and Ziomek also reported discrepancies in the assessment of polarity by SEM and fluorescent labelling. They found that more than 30% of the blastomeres scored as nonpolar by FITC-Con A had microvillous poles under SEM. The actin staining results suggest that microfilaments may play an important role in cell-cell contact and cell flattening in the hamster, as has been shown in the mouse.

In the present study, the microvillous pole in the hamster embryos was found to be restricted to the basal region of cell-cell contact. In the mouse, however, the microvillous pole is located at the outer surface of the embryos. In bovine embryos, the microvillous pole was not always oriented centrally on the apical or the basal regions. The biological significance of the species difference in polarity orientation is not clear.

Another fascinating observation from the present study is that the nucleus of the embryonic blastomere in the hamster manifested a dramatic migration within the cytoplasm during early embryo development. From the early to the late 8-cell stage, hamster embryonic cell nuclei moved from the center to the peripheral, apical region of the cells. To our knowledge, this is the first report describing an outward migration of nuclei in embryos.

In the mouse, Reeve and Kelly reported that during the 4- and 8-cell stages the peripherally located nuclei became clustered nearer the center of the embryo. The direction of nuclear migration seems to be completely reversed in the hamster. Again, the biological significance of the nuclear outward migration and the species difference in the direction of embryonic cell nuclear migration is not known. However, it is apparent that the onset of the observed nuclear migration coincided with the timing of embryo compaction, an important physiological event during early embryo development that leads to subsequent embryonic cell differentiation. The migration of the embryonic cell nuclei is probably caused by rearrangement of cytoskeletal architecture during embryonic compaction. This proposition is supported by the observations that the nuclear position is found in the opposite pole of the polarized microvillous pole in both mice and hamsters (this study). Furthermore, nuclear migration also coincided with the re-location of microtubules and microfilaments during embryo compaction. Additionally, it has been reported that the nucleus (including the germinal vesicle) is located in the microtubule domain of the cells in several species. Polarity of actin distribution was obvious at compaction of the hamster embryos.

It has been demonstrated, by analysis using embedment-free sections, that directional polarity of cytoskeletons occurred in cells from hamster embryos undergoing compaction. In conclusion, embryonic cell polarity in hamsters became manifest during compaction at 8-cell stage; the microvillous pole appeared to be restricted to the basal cell-cell contact region; and an outward migration of the nucleus was observed for the first time in the hamster around the time of embryonic cell compaction.

Mammalian oogenesis involves a highly orchestrated series of mitotic and meiotic cell cycles. In most mammals, most female germ cells undergo a number of mitotic divisions and then enter into meiosis during embryogenesis. In the mouse, oogonia enter prophase of the first meiotic division at embryonic day 13.5 and quickly progress through leptotene, zygotene and pachytene stages before arresting in diplotene, at which point they are surrounded by a single layer of follicle cells.

The oocyte remains arrested in the diplotene stage and pools of resting oocytes are recruited on a cyclical basis to enter a period of oocyte and follicular growth at sexual maturity. At the end of the growth period, the oocyte is meiotically competent, that is, able to undergo the first meiotic division, only to arrest again awaiting fertilization. The progression of the meiotic cell cycle in oogenesis is thus developmentally controlled and the meiotic cell cycle can arrest and resume in response to different stimuli. It is therefore important to understand the role of critical cell cycle-regulatory genes in this process. The cyclins and cyclin-dependent kinases (Cdks) play key roles in the regulation of cellular responses to growth and passage through distinct phases of the cell cycle. Activation of the kinase requires the association of regulatory cyclin subunits, and different cyclins preferentially complex with particular Cdk partners. Regulation of the kinase complexes at distinct cell cycle phases is accomplished in part by differential expression, binding, localization and degradation of cyclins. Among the cyclins, the *in vivo* function of the A-type cyclins has remained somewhat enigmatic.

There are two A-type cyclin proteins in frog, mice and human. The originally described mammalian A-type cyclin, known as cyclin A2, is widely expressed and has been studied extensively both *in vitro* as well as in mammalian cell lines. Cyclin A2 has been shown to be required for S-phase as well as at the G2/M transition and to co-localize with sites of DNA replication in mouse cell lines.

Cyclin A2 can bind and activate both Cdk1 and Cdk2 in somatic cells, consistent with its function in both DNA replication and onset of mitosis. An absolute requirement for cyclin A2 during development and differentiation was elegantly demonstrated by gene targeting. Cyclin A2-deficiency results in embryonic lethality around the peri-implantation stage. Unfortunately, this early lethality precludes studies assessing the effects of lack of cyclin A2 in the tissues and organs of the adult animal.

In contrast to the broad distribution of cyclin A2 expression, cyclin A1 is found at highest levels, if not exclusively, in the adult testis. Although there have been some reports of cyclin A1 expression in other tissues, any such expression is of questionable functional significance given the phenotype of the knockout of the gene encoding mouse cyclin A1. Mutagenesis of *Ccna1* revealed an absolute requirement for progression of male, but not female germ cells through the first meiotic division. Further, Cdk1 forms a complex with cyclin A1 in murine testis and the association of Cdk1 with cyclin A1 is essential for M-phase promoting factor (MPF) activity (cyclin B1/ Cdk1). Given the highly specific expression patterns of the A-type cyclins and essential function of cyclin A1 in the male germ line, we wished to determine their cellular and developmental-specificity of expression in the female germ line and gonad.

Cyclin A2 protein has been shown to be expressed in total ovary and oocytes but the relative levels of mRNA and protein among ovarian cell types and at different stages of oogenesis have not been determined. There also remains some uncertainty with regard to whether cyclin A1 is expressed in female germ cells. This is of particular interest since the frog cyclin A1 is clearly expressed in oocytes and Murphy *et al.*(1997) proposed that maternal cyclin A1 might be involved in enabling early mouse embryos lacking cyclin A2 to undergo a few rounds of cell division.

In the present study we have examined the expression of cyclin A2 at both the mRNA and protein levels during meiotic differentiation of the female germ line. Our findings suggest a role for cyclin A2 in meiosis and mitosis of female germ cells. Furthermore, we have determined the distribution of the cyclin A2-associated proteins Cdk1 and Cdk2 during folliculogenesis in the adult ovary. Finally, we confirm by immunoblot analysis of adult oocytes from normal and cyclin A1-deficient adult mice and by immunohistochemical analysis of embryonic ovaries that *Ccna1* is not expressed in oocytes at stages comparable with its expression in the male germ cells. Thus, cyclin A1 is restricted to the male germ line, consistent with the functional observations resulting from the outcome of gene targeting.

Materials and Methods

Animals and Tissues

Normal tissues for *in situ* hybridization and immunohistochemical analyses were obtained from female C57B16/JF1 mice purchased from The Jackson Laboratory (Bar Harbor, ME, USA). All animals were housed and used in strict adherance to IACUC guidelines.

Ovaries and oviductal ampullae were fixed immediately in 4% paraformaldehyde in 1 x PBS overnight at 4 °C. In some experiments, ampullae were collected from animals that had been superovulated by i.p. injection of 5 IU pregnant mare serum (PMSG) (Sigma), followed 48 h later by injection of 5 IU human chorionic gonadotropin (hCG) (Sigma). The tissues were dissected 15 h after injection with hCG and processed as above. Cyclin A1-deficient females were obtained from the *Ccna1* $^{-/-}$ strain, maintained on a mixed C57Bl/6 x Sv129 background, and genotyped as previously described.

Germinal vesicle (GV)-stage oocytes were collected from ovaries of control and *Ccna1* $^{-/-}$ mice stimulated with PMSG according to our standard procedures. The ovaries were dissected 48 h after PMSG injection and oocytes released by follicle puncture into M2 medium (Special Media, Phillipsburg, NJ, USA) containing 100 μg/ml dibutyryl cAMP (Sigma).

Oocytes at metaphase I were obtained by removing the GV oocytes from the M2 + cAMP medium and culturing them for 15 h in M16 medium (Special Media). Metaphase II-arrested oocytes were collected from mice superovulated as above. Oviducts were dissected 15 h after injection with hCG and oocytes were released into M2 medium. Cumulus cells from oocytes at all stages were dispersed with hyaluronidase and zonae were removed using acid-M2 as described by Simerly & Schatten (1993).

For the embryonic gonad preparations, pregnant female mice from the breeding strain C57Bl6-JF1 or *Ccna1* $^{-/-}$ mice maintained on a mixed Sv129 and C57Bl6 background were killed at daily intervals between 13.5 and 18.5 days after finding a vaginal plug (considered day E0.5). Embryos were removed from the uterus and fixed with 4% paraformaldehyde. The embryonic gonads were identified as ovary or testis by their appearance and location in the embryo, under a dissecting microscope. The embryonic ovary was removed and embedded in 1% agarose and then embedded in paraffin using our previously described standard procedures. A portion of the embryonic liver was used as a source of DNA to confirm the genotype of the embryos (*Ccna1* $^{-/-}$ *Ccna1* $^{+/-}$ or *Ccna1* $^{+/+}$).

Probes and Reagents

Probes corresponding to the 5' end of the murine cyclin A2 cDNA were generated from plasmid p4-2 0.4. Sense and antisense RNA probes for *in situ* hybridization were transcribed from the T7 and T3 promoters of linearized plasmids using the appropriate RNA polymerase, following the manufacturer's suggested protocols (Promega). Probes used for *in situ* hybridization were labeled with ^{35}S-labeled UTP (Dupont) and hydrolyzed to approximately 0.5 kb in size. Three sources of antibodies against cyclin A2 were used to verify the expression and subcellular localization of cyclin A2 in embryonic and adult stages of ovary. One was kindly provided by Dr Mark Carrington (Cambridge University, UK) and used at a dilution of 1:500. Two other antibodies were purchased from Upstate Biotechnology Inc. (Lake Placid, NY, USA) and Oncogene Sciences (Cambridge, MA, USA) and were used at dilutions of 1:250 or 1:500 respectively. Cdk1 and Cdk2 antibodies were purchased from Upstate Biotechnology and were used at a dilution of 1:500. Polyclonal antibody against cyclin A1 was produced in our laboratory according to procedures previously described and was used at a dilution of 1:250.

In Situ Hybridization

Paraformaldehyde-fixed tissues were embedded in paraffin, cut into 6 μm sections and analyzed by *in situ* hybridization using our previously described protocols. After hybridization, slides were washed at a final stringency of 0.1 x SSC at 65 °C for 2 h. Autoradiography was performed for 2 weeks. Developed slides were counterstained with hematoxylin and eosin (H&E) and viewed on a Leitz photomicroscope under brightfield and epiluminescence optics. Photomicrographs were taken using Kodak Ektachrome 64T film.

Immunohistochemistry

Paraformaldehyde-fixed, paraffin-embedded tissues were sectioned at 6 μm and analyzed by immunohistochemistry using the Vectastain ABC kit (Vector Laboratories, Burlingame, CA, USA). Tissue sections were deparaffinized in xylene, hydrated through an alcohol series, and washed with water, then boiled in antigen-retrieval buffer (0.01 M citrate–acetate, pH 6.0) in a microwave for 10 min and washed extensively with distilled water. The slides were treated with 0.03% H_2O_2 in methanol for 20 min at 4 °C, washed with 1 x PBS with 0.1% Triton X-100 (PBSTr) and blocked for 1 h with 2.5% goat serum in PBSTr. Incubation with primary antibodies was carried out overnight at 4 °C in a humidified chamber. The slides were washed three times, 10 min each, with PBSTr and processed using the Vectastain ABC kit, following the manufacturer's suggestions. The 3,3'-diaminobenzidine (DAB)-stained slides were counterstained with hematoxylin and viewed on a Nikon 800 photomicroscope under brightfield optics.

Immunoblotting

Fresh or frozen tissue or oocytes were lysed in lysis buffer (150 mM NaCl, 50 mM Tris–HCl pH 7.5, 1% NP-40, 10 mM NaF, 10 mM phenylmethylsulfonyl fluoride and the protease inhibitors). Lysates were cleared by centrifugation at 5000 *g* for 10 min at 4 °C. Equal amounts of proteins from each sample were then subjected to electrophoresis on a 12% SDS-PAGE gel followed by transfer to Hybond enhanced chemiluminescence (ECL) detection nitrocellulose membrane (Amersham Pharmacia Biotech). The membranes were blocked in 6% no-fat milk in Tris-buffered saline overnight and then incubated with primary antibodies (cyclin A1 at a 1:1000 dilution) for 1–2 h, followed by horseradish peroxidase-conjugated secondary antibodies (Boehringer-Mannheim, Indianapolis, IN, USA) and visualized using the ECL detection system and ECL films (Amersham Pharmacia Biotech).

Results

Cyclin A2 is expressed in distinct cell types in the adult ovary: *In situ* hybridization analysis and immunohistochemical methods were used to determine the cellular sites of *Ccna2* expression in the adult ovary and in isolated oocytes. Oocytes in the adult ovary, which are arrested in diplotene of prophase I, can be staged according to their size and the morphological characteristics of the surrounding follicle cells. *In situ* hybridization results showed that while *Ccna2* transcripts were detected in oocytes and granulosa cells at all stages of follicular development, the signals were relatively weak in the granulosa cells in early-stage follicles. Much higher levels of signal were observed in the granulosa cells of growing follicles, particularly in the cumulus layer (cells immediately surrounding the oocyte) in stage 6–8 follicles. The levels of *Ccna2* hybridization signal present over the oocytes remained constant in oocytes of stage 6 and 7 follicles and in ovulated eggs within the oviductal ampulla. In contrast to the readily detectable *Ccna2* transcripts, we were unable to detect transcripts for *Ccna1* at this level of sensitivity.

To determine whether cyclin A2 expression is regulated at the protein level, we performed immunohistochemistry on ovary sections containing various stages of follicular development. The cellular distribution of cyclin A2 protein appeared to be similar to that observed for *Ccna2* transcripts. In smaller follicles (type 1 and 2), cyclin A2 was clearly detected in the oocyte and at low levels if at all in the follicle cells. In type 3 and 4 follicles, cyclin A2 continued to be expressed in the oocytes but now also in granulosa cells. As follicular growth proceeded, more granulosa cells expressed cyclin A2 protein and at higher levels, while the staining intensity in the oocytes remained approximately the same in the now type 5a and type 5b follicles. It is of interest to note that in large (type 6) follicles and mature type 8 follicles, cyclin A2 protein was intensely stained in the granulosa cells and was particularly obvious in the cumulus. The levels of cyclin A2 expression in the oocytes of stage 8 follicles appeared to be much lower than those at earlier follicle stages. We also noted that some thecal cells surrounding the follicles were positive for cyclin A2 staining whereas no staining was detected in stromal cells. No staining was observed when control, pre-immune serum was used.

We extended this analysis to include ovaries from mice that were pre-pubertal (post-natal days 7 and 14) along with post-pubertal young adult (30 and 60 days) to determine the stage of growth and differentiation of the follicle was the primary regulator of cyclin A2 expression or if the hormonal status of animals at pre- or post-pubertal ages might also play a role.

The relative levels of cyclin A2 protein in oocytes, granulosa cells and oocyte–cumulus complexes in ovaries collected at the time periods noted above were evaluated and rated from + to +++. The relative staining intensity for cyclin A2 in oocytes remained approximately the same in all of the follicles in pre-pubertal ovaries, which contained follicles up to early stage 6. In the post-pubertal ovaries, levels of cyclin A2 were also more abundant in the oocytes in immature follicles but as the follicles reached stages 6–8, the staining intensity dropped noticeably. In the granulosa cells, the patterns of cyclin A2 expression showed differentiation stage-dependent increases with follicular development in both pre-pubertal and post-pubertal ovaries. Relatively high levels of cyclin A2 were detected in the cumulus. The levels of cyclin A2 expression in granulosa cells correlated positively with the stage of follicular growth and differentiation, while in the oocyte, more mature oocytes contained less cyclin A2, regardless of the hormonal state.

Localization of cyclin A2's Cdk partners in distinct cell types: We next determined if the putative Cdk partners for cyclin A2, Cdk1 and Cdk2, were expressed in the corresponding cell types in the adult ovary. Immunohistochemical analysis revealed the expression of Cdk1 protein in oocytes and some follicle cells of type 2 and type 3 follicles and its absence in stromal cells.

In type 4, 5a and 5b follicles, more follicle cells showed expression of Cdk1 and it was present in the oocyte–cumulus complexes and within the oocyte of type 6 and 8 follicles. Occasionally, there was a non-specific staining that localized to the zona pellucida of larger oocytes, but this staining was highly variable. Similar to what was observed for cyclin A2, the staining intensity was greater in follicle cells as compared with the oocytes. However, no apparent increase in levels of Cdk1 expression was detected in follicular cells of type 8 follicles compared with the earlier stages, which was in contrast to what was observed for cyclin A2. The levels of Cdk1 expression also appeared to be similar in oocytes of different follicle stages.

Placenta

The placenta is an organ that connects the developing fetus to the uterine wall to allow nutrient uptake, waste elimination and gas exchange via the mother's blood supply. Placentas are a defining characteristic of eutherian or "placental" mammals, but are also found in some snakes and lizards with varying levels of development up to mammalian levels.

Protherial (egg-laying) and metatherial (marsupial) mammals produce a choriovitelline placenta that, while connected to the uterine wall, provides nutrients mainly derived from the egg sac. The placenta develops from the same sperm and egg cells that form the fetus, and functions as a fetomaternal organ with two components, the fetal part (Chorion frondosum), and the maternal part (Decidua basalis).

Structure

In humans, the placenta averages 22 cm (9 inch) in length and 2–2.5 cm (0.8–1 inch) in thickness (greatest thickness at the center and become thinner peripherally). It typically weighs approximately 500 grams (1 lb). It has a dark reddish-blue or maroon colour. It connects to the fetus by an umbilical cord of approximately 55–60 cm (22–24 inch) in length that contains two arteries and one vein. The umbilical cord inserts into the chorionic plate (has an eccentiric attachment).

Vessels branch out over the surface of the placenta and further divide to form a network covered by a thin layer of cells. This results in the formation of villous tree structures. On the maternal side, these villous tree structures are grouped into lobules called cotyledons. In humans the placenta usually has a disc shape but different mammalian species have widely varying shapes.

Development

The placenta begins to develop upon implantation of the blastocyst into the maternal endometrium. The outer layer of the blastocyst becomes the trophoblast which forms the outer layer of the placenta. This outer layer is divided into two further layers: the underlying cytotrophoblast layer and the overlying syncytiotrophoblast layer.

The syncytiotrophoblast is a multinucleate continuous cell layer which covers the surface of the placenta. It forms as a result of differentiation and fusion of the underlying cytotrophoblast cells, a process which continues throughout placental development.

The syncytiotrophoblast (otherwise known as syncytium), thereby contributes to the barrier function of the placenta. The placenta grows throughout pregnancy. Development of the maternal blood supply to the placenta is suggested to be complete by the end of the first trimester of pregnancy (approximately 12–13 weeks).

Placental Circulation

Maternal Placental Circulation

In preparation for implantation, the uterine endometrium undergoes 'decidualisation'. Spiral arteries in the decidua are remodelled so that they become less convoluted and their diameter is increased. This increases maternal blood flow to the placenta and also decreases resistance so that shear stress is reduced. The relatively high pressure as the maternal blood enters the intervillous space through these spiral arteries bathes the villi in blood. An exchange of gases takes place. As the pressure decreases, the deoxygenated blood flows back through the endometrial veins.

Fetoplacental Circulation

Deoxygenated fetal blood passes through umbilical arteries to the placenta. At the junction of umbilical cord and placenta, the umbilical arteries branch radially to form chorionic arteries. Chorionic arteries also branch before they enter into the villi. In the villi, they form an extensive arteriocapillary venous system, bringing the fetal blood extremely close to the maternal blood; but no intermingling of fetal and maternal blood occurs.

Functions

Nutrition: The perfusion of the intervillous spaces of the placenta with maternal blood allows the transfer of nutrients and oxygen from the mother to the fetus and the transfer of waste products and carbon dioxide back from the fetus to the mother. Nutrient transfer to the fetus is both actively and passively mediated by proteins called nutrient transporters that are expressed within placental cells.

Adverse pregnancy situations, such as those involving maternal diabetes or obesity, can increase or decrease levels of nutrient transporters in the placenta resulting in overgrowth or restricted growth of the fetus.

Metabolic and Endocrine Activity

In addition to the transfer of gases and nutrients, the placenta also has metabolic and endocrine activity. It produces, among other hormones, progesterone, which is important in maintaining the pregnancy; somatomammotropin (also known as placental lactogen), which acts to increase the amount of glucose and lipids in the maternal blood; estrogen; relaxin, and beta human chorionic gonadotrophin (beta-hCG). This results in increased transfer of these nutrients to the fetus and is also the main cause of the increased blood sugar levels seen in pregnancy. This hormone (beta-hCG) ensures that progesterone and oestrogen are secreted; progesterone and oestrogen thicken and maintain the uterine lining as well as inhibit the production and release of more eggs. However after about 2 months the placenta takes on the role of producing progesterone and therefore beta-hCG is no longer needed. Beta-hCG is excreted in urine and this is what pregnancy tests detect. It also produces insulin-like growth factors (IGFs).

Cloaking from Immune System of Mother

The placenta and fetus may be regarded as a foreign allograft inside the mother, and thus must evade from attack by the mother's immune system.

For this purpose, the placenta uses several mechanisms:

- It secretes Neurokinin B containing phosphocholine molecules. This is the same mechanism used by parasitic nematodes to avoid detection by the immune system of their host.
- Also, there is presence of small lymphocytic suppressor cells in the fetus that inhibit maternal cytotoxic T cells by inhibiting the response to interleukin 2.

However, the placental barrier is not the sole means to evade the immune system, as foreign fetal cells also persist in the maternal circulation, on the other side of the placental barrier.

Birth

When the fetus is born, its placenta begins a physiological separation for spontaneous expulsion afterwards (and for this reason is also called the afterbirth). In humans, the umbilical cord is routinely clamped and severed prior to the delivery of the placenta, often within seconds or minutes of birth, a medical protocol known as 'active management of third stage' which has been called into question by advocates of natural birth and 'passive management of third stage' The site of the former umbilical cord attachment in the center of the front of the abdomen is known as the navel, umbilicus, or belly-button.

Modern obstetric practice has decreased maternal death rates enormously. The addition of active management of the third stage of labour is a major contributor towards this. It involves giving oxytocin via IM injection, followed by cord traction to assist in delivering the placenta. Premature cord traction can pull the placenta before it has naturally detached from the uterine wall, resulting in hemorrhage. The BMJ summarized the Cochrane group metanalysis (2000) of the benefits of active third stage as follows:

"One systematic review found that active management of the third stage of labour, consisting of controlled cord traction, early cord clamping plus drainage, and a prophylactic oxytocic agent, reduced postpartum haemorrhage of 500 or 1000 mL or greater and related morbidities including mean blood loss, postpartum haemoglobin less than 9 g/dL, blood transfusion, need for supplemental iron postpartum, and length of third stage of labour. Although active management increased adverse effects such as nausea, vomiting, and headache, one RCT identified by the review found that women were less likely to be dissatisfied when their third stage of labour was actively managed."

Risks of retained placenta include hemorrhage and infection. If the placenta fails to deliver in 30 minutes in a hospital environment, manual extraction may be required if heavy ongoing bleeding occurs, and very rarely a curettage is necessary to ensure that no remnants of the placenta remain (in rare conditions with very adherent placenta (placenta accreta)). However, in birth centers and attended home birth environments, it is common for licensed care providers to wait for the placenta's birth up to 2 hours in some instances.

Non-humans

In most mammalian species, the mother bites through the cord and consumes the placenta, primarily for the benefit of prostaglandin on the uterus after birth. This is known as placentophagy. However, it has been observed in zoology that chimpanzees, with which humans share 94%-99% of genetic material, apply themselves to nurturing their offspring, and keep the fetus, cord, and placenta intact until the cord dries and detaches the next day.

Top: Human placenta shown a few minutes after birth. The side shown faces the baby with the umbilical cord top right. The white fringe surrounding the bottom is the remnants of the amniotic sac. *Bottom:* A different placenta displays side that connects to the uterine wall. The placenta exists in most mammals and some reptiles. It is probably polyphyletic.

Pathology

Numerous pathologies can affect the placenta.

When the placenta implants too deeply:

- Placenta accreta
- Placenta praevia
- Placental abruption

Infections involving the placenta:

- Placentitis, such as the TORCH infections.
- Chorioamnionitis.

Use in Medicine

Human placenta is increasingly being used in western medicine with even a branch called "placenta pharmacology" being updated regularly. It has been proven to be effective in treating sores and making tissues soft and heal better.

Cultural Practices and Beliefs

The placenta often plays an important role in various human cultures, with many societies conducting rituals regarding its disposal. In the Western world, the placenta is most often incinerated. Some cultures bury the placenta for various reasons. The Mâori of New Zealand traditionally bury the placenta from a newborn child to emphasize the relationship between humans and the earth. Similarly, the Navajo bury the placenta and umbilical cord at a specially chosen site, particularly if the baby dies during birth. In Cambodia and Costa Rica, burial of the placenta is believed to protect and ensure the health of the baby and the mother. If a mother dies in childbirth, the Aymara of Bolivia bury the placenta in a secret place so that the mother's spirit will not return to claim her baby's life. The placenta is believed by some communities to have power over the lives of the baby or its parents. The Kwakiutl of British Columbia bury girls' placentas to give the girl skill in digging clams, and expose boys' placentas to ravens to encourage future prophetic visions. In Turkey, the proper disposal of the placenta and umbilical cord is believed to promote devoutness in the child later in life. In Ukraine, Transylvania, and Japan, interaction with a disposed placenta is thought to influence the parents' future fertility. Several cultures believe the placenta to be or have been alive, often a relative of the baby. Nepalese think of the placenta as a friend of the baby's; Malaysian Orang Asli regard it as the baby's older sibling. The Ibo of Nigeria consider the placenta the deceased twin of the baby, and conduct full funeral rites for it. Native Hawaiians believe that the placenta is a part of the baby, and traditionally plant it with a tree which can then grow alongside the child. In some cultures, the placenta is eaten, a practice known as placentophagy. In some eastern cultures, such as China and Hong Kong, the placenta is thought to be healthful and is used in medicine and various health products. In Western cultures, the consumption of humans or parts thereof by other humans is seen as anthropophagy and may be subject to legal prosecution. However, recent research shows many benefits from placentophagy, and across the nation, this practice is gaining popularity.

Eutheria

Eutheria are a group of mammals consisting of placental mammals plus all extinct mammals that are more closely related to living placentals (such as humans) than to living marsupials (such as kangaroos). They are distinguished from non-eutherians by various features of the feet, ankles, jaws and teeth. One of the major differences between placental and non-placental eutherians is that placentals lack epipubic bones, which are present in all other fossil and living mammals.

The earliest known fossil eutherian, *Eomaia* was found in Asia and is dated to the Early Cretaceous period, about 125 million years ago.

Definition

Eutherians are a group of mammals consisting of placental mammals plus all extinct mammals that are more closely related to living placentals (such as humans) than to living marsupials (such as kangaroos). There are no living non-placental eutherians, and so knowledge of their synapomorphies ("defining features") is entirely based on a few fossils – which means the reproductive features that distinguish modern placentals from other mammals cannot be used in defining the eutheria. The features of eutheria that distinguish them from metatherians, a group that includes modern marsupials, are:

- an enlarged malleolus ("little hammer") at the bottom of the tibia, the larger of the two shin bones.
- the joint between the first metatarsal bone and the entocuneiform bone in the foot is offset further back than the joint between the second metatarsal and mesocuneiform bones – in metatherians these joints are level with each other.
- various features of jaws and teeth.

Reproductive features are also of no use in identifying fossil placental mammals, which are distinguished from other eutherians by:

- the presence of a malleolus at the bottom of the fibula, the smaller of the two shin bones.
- a complete mortise and tenon upper ankle joint, where the rearmost bones of the foot fit into a socket formed by the ends of the tibia and fibula.
- a wide opening at the bottom of the pelvis, which allows the birth of large, well-developed offspring. Marsupials have and non-placental eutherians had a narrower opening that allows only small, immature offspring to pass through.

- the absence of epipubic bones extending forwards from the pelvis, which are not found in any placental, but are found in all other mammals – non-placental eutherians, marsupials, monotremes and mammaliformes– and even in the cynodont therapsids that are closest to mammals. Their function is to stiffen the body during locomotion. This stiffening would be harmful in pregnant placentals, whose abdomens need to expand.

Subgroups

These are the subgroups of extant members of Eutheria:

- Euarchontoglires, e.g. monkeys, rats, hares
- Laurasiatheria, e.g. cattle, whales, bats, cats
- Xenarthra, e.g. armadillos, anteaters
- Afrotheria, e.g. elephants, manatees.

These groups together make up the crown group Placentalia (placental mammals). Eutheria also includes now extinct lineages that lie outside of Placentalia.

Prototheria

Prototheria is a taxonomic group, or taxon, to which the order Monotremata belongs. It is conventionally ranked as a subclass within the mammals.

Most of the animals in this group are extinct. The egg-laying monotremes are known from fossils of the Cretaceous and Cenozoic periods; they are represented today by the platypus and several species of echidna.

The names Prototheria, Metatheria and Eutheria (meaning "first beasts", "changed beasts", and "true beasts") refer to the three mammalian groupings which have living representatives. Each of the three may be defined as a putative clade comprising a living crown-group (respectively the Monotremata, Marsupialia and Placentalia) plus any fossil species which are more closely related to that crown-group than to any other animals. The threefold division of living mammals into monotremes, marsupials and placentals was already well established when Thomas Huxley proposed the names Metatheria and Eutheria to incorporate the two latter groups in 1880. Initially treated as subclasses, Metatheria and Eutheria are by convention now grouped as infraclasses of the subclass Theria, and in more recent proposals have been demoted further (to cohorts or even magnorders), as cladistic reappraisals of the relationships between living and fossil mammals have suggested that the Theria itself should be reduced in rank. Prototheria, on the other hand, was generally recognised as a subclass until quite recently, on the basis of an hypothesis which defined the group by two supposed synapomorphies: (1) formation of the side wall of the braincase from a bone called the anterior lamina, contrasting with the alisphenoid in therians; and (2) a linear alignment of molar cusps, contrasting with a triangular arrangement in therians. These characters appeared to unite monotremes with a range of Mesozoic fossil orders in a broader clade for which the name Prototheria was retained, and of which monotremes were thought to be only the last surviving branch.

The evidence which was held to support this grouping is now universally discounted. In the first place, examination of embryos has revealed that the development of the braincase wall is essentially identical in therians and in 'prototherians': the anterior lamina simply fuses with the alisphenoid in therians, and therefore the 'prototherian' condition of the braincase wall is primitive for all mammals while the therian condition can be derived from it. Additionally, the linear alignment of molar cusps is also primitive for all mammals. Therefore, neither of these states can supply a uniquely shared derived character which would support a 'prototherian' grouping of orders in contradistinction to Theria.

In a further reappraisal, the molars of embryonic and fossil monotremes (living monotreme adults are toothless) appear to demonstrate an ancestral pattern of cusps which is similar to the triangular arrangement observed in therians. Some peculiarities of this dentition support an alternative grouping of monotremes with certain recently-discovered fossil forms into a proposed new clade known as the Australosphenida, and also suggest that the triangular array of cusps may have evolved independently in australosphenidans and therians.

The Australosphenida hypothesis remains controversial, and some taxonomists prefer to maintain the name Prototheria as a fitting contrast to the other group of living mammals, the Theria. In theory, the Prototheria is taxonomically redundant, since Monotremata is currently the only order which can still be confidently included, but its retention might be justified if new fossil evidence, or a re-examination of known fossils, enables extinct relatives of the monotremes to be identified and placed within a wider grouping.

Placentation in Mammals

In placental mammals, the placenta forms after the embryo implants into the wall of the uterus. The developing fetus is connected to it via an umbilical cord. Animal placentas are classified based on the number of tissues separating the maternal from the fetal blood. The placentation types found in animals are:

- endotheliochorial placentation (e.g. in most carnivores like cats and dogs)
- epitheliochorial placentation (e.g. in ruminants, horses, whales)
- hemochorial placentation (e.g. in most primates, including humans)

During pregnancy, placentation is the formation and growth of the placenta inside the uterus. It occurs after the implantation of the embryo into the uterine wall and involves the remodeling of blood vessels in order to supply the needed amount of blood. In humans, placentation takes place 7–8 days after fertilization.

In the case of twins, dichorionic placentation refers to the presence of two placentas (in all dizygotic and some monozygotic twins). Monochorionic placentation occurs when monozygotic twins develop with only one placenta and bears a higher risk of complications during pregnancy. Abnormal placentation can lead to an early termination of pregnancy, for example in pre-eclampsia.

Placenta Accreta

Placenta accreta is a severe obstetric complication involving an abnormally deep attachment of the placenta, through the endometrium and into the myometrium (the middle layer of the uterine wall). There are three forms of placenta accreta, distinguishable by the depth of penetration. The placenta usually detaches from the uterine wall relatively easily, but women who encounter placenta accreta during childbirth are at great risk of haemorrhage during its removal. This commonly requires surgery to stem the bleeding and fully remove the placenta, and in severe forms can often lead to a hysterectomy or be fatal.

Placenta accreta affects approximately 1 in 2,500 pregnancies.

Variants

The most common form of placenta accreta is an invasion of the myometrium which does not penetrate the entire thickness of the muscle. This form of the condition accounts for around 75-78% of all cases, and has no name other than placenta accreta. There are two further variants of the condition that are known by specific names and are defined by the depth of their attachment to uterine wall. Placenta increta occurs when the placenta further extends into the myometrium and happens in around 17% of all cases.

Placenta percreta, the worst form of the condition and occurring in 5-7% of cases, is when the placenta penetrates the entire myometrium to the uterine serosa (invades through entire uterine wall). This variant can lead to the placenta attaching to other organs such as the rectum or bladder.

Diagnosis

Placenta accreta is very rarely recognised before birth, and is very difficult to diagnose. An ultrasound can lead to the diagnosis of a suspected accreta and an MRI will give more detail leading to further suspicion of such an abnormal placenta.

However, both the ultrasound as the MRI rarely confirm an accreta with certainty. While it can lead to some vaginal bleeding during the third trimester, this is more commonly associated with the factors leading to the condition.

In some cases the second trimester can see elevated maternal serum alpha-fetoprotein levels, though this is also an indicator of many other conditions.

Risk Factors

The condition affects around 10% of cases of placenta praevia, and is increased in incidence by the presence of scar tissue i.e.. Asherman's syndrome usually from past uterine surgery, especially from a past D&C, (which is used for many indications including miscarriage, termination, and postpartum hemorrhaging), myomectomy, or caesarean section. A thin decidua can also be a contributing factor to such trophoblastic invasion. Some studies suggest that the rate of incidence is higher when the fetus is female.

Placenta Praevia

Placenta praevia (placenta previa AE) is an obstetric complication in which the placenta is attached to the uterine wall close to or covering the cervix. It can sometimes occur in the later part of the first trimester, but usually during the second or third. It is a leading cause of antepartum haemorrhage (vaginal bleeding). It affects approximately 0.5% of all labours.

Pathophysiology

No specific cause of placenta previa has yet been found but it is hypothesized to be related to abnormal vascularisation of the endometrium caused by scarring or atrophy from previous trauma, surgery, or infection. In the last trimester of pregnancy the isthmus of the uterus unfolds and forms the lower segment. In a normal pregnancy the placenta does not overlie it, so there is no bleeding. If the placenta does overlie the lower segment, it may shear off and a small section may bleed.

Women with placenta previa often present with painless, bright red vaginal bleeding. This bleeding often starts mildly and may increase as the area of placental separation increases.

Praevia should be suspected if there is bleeding after 24 weeks of gestation. Abdominal examination usually finds the uterus non-tender and relaxed. Leopold's Maneuvers may find the fetus in an oblique or breech position or lying transverse as a result of the abnormal position of the placenta. Praevia can be confirmed with an ultrasound. In parts of the world where ultrasound is unavailable, it is not uncommon to confirm the diagnosis with an examination in the surgical theatre.

The proper timing of an examination in theatre is important. If the woman is not bleeding severely she can be managed non-operatively until the 36th week. By this time the baby's chance of survival is as good as at full term.

Placenta previa is classified according to the placement of the placenta:

- Type I or low lying: The placenta encroaches the lower segment of the uterus but does not infringe on the cervical os.
- Type II or marginal: The placenta touches, but does not cover, the top of the cervix.
- Type III or partial: The placenta partially covers the top of the cervix
- Type IV or complete: The placenta completely covers the top of the cervix.

Placenta previa is itself a risk factor of placenta accreta.

Risk Factors

The following have been identified as risk factors for placenta praevia:

- Previous placenta previa, caesarean delivery, or abortion.
- Women who have had previous pregnancies, especially a large number of closely spaced pregnancies, are at higher risk.
- Women who are younger than 20 are at higher risk and women older than 30 are at increasing risk as they get older.
- Women with a large placentae from twins or erythroblastosis are at higher risk.
- Women who smoke or use cocaine may be at higher risk.
- Race is a controversial risk factor, with some studies finding that people from Asia and Africa are at higher risk and others finding no difference.

Intervention

An initial assessment to determine the status of the mother and fetus is required. Although mothers used to be treated in the hospital from the first bleeding episode until birth, it is now considered safe to treat placenta praevia on an outpatient basis if the fetus is at less than 30 weeks of gestation, and neither the mother nor the fetus are in distress.

Immediate delivery of the fetus may be indicated if the fetus is mature or if the fetus or mother are in distress. Blood volume replacement (to maintain blood pressure) and blood plasma replacement (to maintain fibrinogen levels) may be necessary.

It is controversial if vaginal delivery or a Caesarean section is the safest method of delivery. In cases of fetal distress a Caesarean section is indicated. Caesarian section is contraindicated in cases of disseminated intravascular coagulation.

A problem exists in places where a Caesarean section cannot be performed, due to the lack of a surgeon or equipment. In these cases the infant can be delivered vaginally. There are two ways of doing this with a placenta praevia:

- The baby's head can be brought down to the placental site (if necessary with Willet's forceps or a vulsellum) and a weight attached to his scalp
- A leg can be brought down and the baby's buttocks used to compress the placental site.

The goal of this type of delivery is to save the mother, and both methods will often kill the baby. These methods were used for many years before Caesarean section and saved the lives of both mothers and babies with this condition.

The main risk with a vaginal delivery with a praevia is that as you are trying to bring down the head or a leg, you might separate more of the placenta and increase the bleeding.

Placenta praevia increases the risk of puerperal sepsis and postpartum haemorrhage because the lower segment to which the placenta was attached contracts less well post-delivery.

6

Mammal Anatomy

Abomasum

The abomasum, also known as the maw, and the rennet-bag, and the read, is the fourth and final stomach compartment in ruminants. It secretes rennet, an ingredient used in cheese creation. The word *abomasum* is from New Latin and it was first used in English in 1706. It comes from Latin *ab-* + *omasum* "intestine of an ox," and it is possibly from the Gaulish language.

The abomasum's normal anatomical location is along ventral midline. It is a secretory stomach similar in anatomy and function as the monogastric stomach. It serves primarily in the acid hydrolysis of microbial and dietary protein, preparing these protein sources for further digestion and absorption in the small intestine.

Dairy cattle on high production diets are susceptible to a number of pathologies, most commonly after calving. A gas filled abomasum can move into an abnormal location resulting in left displaced abomasum (LDA) or right displaced abomasum (RDA). If the abomasum displaces to the right, it is at risk of torsion and becoming a right torsioned abomasum (RTA).

A displaced abomasum will cause cows to present all or some of the following signs: loss of appetite, decrease rumen contractions, decrease cud chewing, and drop in milk production. While an LDA and RDA are not immediately life threatening, veterinary care is required for surgical correction.

Abomasitis is a relatively rare, but serious, disease of the abomsum whose causes are currently unknown.

Antler

Antlers are the usually large and complex bony appendages on the heads of most deer species, mostly on males; only caribou and reindeer have antlers on the females, and these are normally smaller than those of the males. Nevertheless, fertile does from other species of deer have the capacity to produce antlers on occasion, usually due to increased testosterone levels.

Each antler grows from an attachment point on the skull called a pedicle. While an antler is growing, it is covered with highly vascular skin called velvet, which supplies oxygen and nutrients to the growing bone; once the antler has achieved its full size, the velvet is lost and the antler's bone dies. This dead bone structure is the mature antler, which is itself shed after each mating season.

An advancer is a term used by hunters for one of the starts or branches of a buck's attire between the back antler and the "palm" (the flattened part of the antlers). In other words, the advancer is the second branch of a buck's horn.

Antlers appear to act as large hearing aids. This effect was discovered by researchers George and Peter Bubenik and published the findings in the *European Journal of Wildlife* in March 2008. Moose with antlers have far more sensitive hearing than moose without. The Bubeniks then studied trophy antlers with an artificial ear, confirming that the antler behaves like a parabolic reflector.

Uses

Antlers that are shed each year fall to the ground and are often collected by craftsmen. Called "fallen", they have been used for centuries by mankind to make tools, toys and games. Collected while in the velvet stage, antlers of elk and deer are commonly used as a dietary supplement or alternative medicinal substance in Asia; the practice dates to more than 2,000 years ago.

Calcar

The calcar, also known as the calcaneum, is the name given to a spur of cartilage arising from inner side of ankle and running along part of outer interfemoral membrane in bats, this is to help spread the interfemoral membrane, which is part of the wing membrane between the tail and the hind legs.

This membrane can form a basket or pouch to help catch and hold insects captured in flight . The oldest known ancestor to present day bats, Icaronycteris index, apparently did not have a calcar or spur as evidenced by fossil remains. Calcar (Femorale) also refers to the dense, vertically orientated bone present in the posteroemedial region of the femoral shaft under the lesser trochanter of the femur.

Canine Tooth

In mammalian oral anatomy, the canine teeth, also called cuspids, dogteeth, fangs, or (in the case of those of the upper jaw) eye teeth, are relatively long, pointed teeth. However, they can appear more flattened, causing them to resemble incisors and leading them to be called *incisiform*. They evolved and are used primarily for firmly holding food in order to tear it apart, and occasionally as weapons. They are often the largest teeth in a mammal's mouth. Most species that develop them normally have four per individual, two in the upper jaw and two in the lower, separated within each jaw by its incisors; humans and dogs are examples. In most animals, canines are the anterior-most teeth in the maxillary bone.

The four canines in humans are the two maxillary canines and the two mandibular canines.

Details

There are four canine teeth: two in the upper (maxillary) and two in the lower (mandibular) arch. A canine is placed laterally to each lateral incisor. They are larger and stronger than the incisors, and their roots sink deeply into the bones, and cause well-marked prominences upon the surface. The crown is large and conical, very convex on its labial surface, a little hollowed and uneven on its lingual surface, and tapering to a blunted point or cusp, which projects beyond the level of the other teeth. The root is single, but longer and thicker than that of the incisors, conical in form, compressed laterally, and marked by a slight groove on each side.

In humans, the upper canine teeth (popularly called eye teeth, from their position under the eyes) are larger and longer than the lower, and usually present a distinct basal ridge. Eruption typically occurs between the ages of 11 and 12 years of age.

From a facial aspect, maxillary canines are approximately 1mm narrower than the central incisor. Their mesial aspects resemble the adjacent lateral incisors, while their distal aspects anticipate the first premolars. They are slightly darker and more yellow in colour than the other anterior teeth. From a lingual aspect, they have well developed mesial and distal marginal ridges and a well-developed cingulum. A prominent lingual ridge divides the lingual aspect in half and creates the mesial and distal lingual fossae between the lingual ridge and the marginal ridges. From a proximal aspect, they resemble the incisors, but are more robust, especially in the cingulum region. Incisally, they are visibly asymmetrical, as the mesial incisal edge is slightly shorter than the distal incisal edge, which places the cusp slightly mesial to the long axis of the tooth.

They are also thicker labiolingually than mesio-distally. Because of the disproportionate incisal edges, the contacts are also asymmetrical. Mesially, the contact sits at the junction of the incisal and middle third of the crown, while distally, the contact as more cervical, in the middle of the middle third of the crown.

The lower canine teeth (popularly called stomach teeth) are placed nearer the middle line than the upper, so that their summits correspond to the intervals between the upper canines and the lateral incisors. Eruption typically occurs between the ages of 9 and 10 years of age. From a facial aspect, the mandibular canine is noticeably narrower mesio-distally than the maxillary one, even though the root may be just as long, if not longer (and sometimes bifurcated). A distinctive feature is the nearly straight outline this tooth has compared to the maxillary canine which is slightly more bowed. As in the maxillary canine, the mesial incisal edge (or cusp ridge) is shorter than the distal side, however, the cusp is displaced slightly lingual relative to the cusp of the maxillary canine. Lingually, the surface of the tooth is much more smooth compared to the very pronounced surface of the maxillary canine, and the cingulum is noted as less developed.

Sexual Dimorphism

With many animals the canine teeth in the upper or lower jaw, or in both, are much larger in the males than in the females, or are absent in females, except sometimes a hidden rudiment. Certain antelopes, the musk-deer, camel, horse, boar, various apes, seals, and the walrus, offer instances.

In Non-mammals

In non-mammals, teeth similar to canines may be termed "caniniform" ("canine-shaped") teeth.

Caudofemoralis

Location: The caudofemoralis is found directly posterior (caudal) to the Gluteus maximus and directly anterior (cranial) to the Biceps femoris.

Origin (Anatomy) and Insertion (Anatomy)

The Caudofemoralis originates from the transverse processes of the second, third and fourth caudal vertebrae. The caudodistal portion of the muscle lies deep to the proximocranial portion of the Biceps femoris. Near the middle of the thigh, the Caudofemoralis gives rise to a long, thin, and narrow tendon that passes distally to the knee joint and inserts into the fascia lata that is anchored to the lateral border of the patella.

Action

The Caudofemoralis acts to flex the tail laterally to its respective side when the pelvic limb is bearing weight. When the pelvic limb is lifted off the ground, contraction of the Caudofemoralis causes the limb to abduct and the shank to extend by extending the hip joint (acetabulofemoral or coxofemoral joint).

Cloven Hoof

A cloven hoof is a hoof split into two toes. This is found on members within the mammalian order Artiodactyla. Examples of mammals that possess this type of hoof are pigs, deer and sheep. The two digits of cloven hoofed animals are homologous to the third and fourth fingers of the hand. They are called claws and are named for their relative location on the foot: the outer, or lateral, claw and the inner, or medial claw. The space between the two claws is called the interdigital cleft; the area of skin is called the interdigital skin. The hard outer covering of the hoof is called the hoof wall, or horn. It is a hard surface, similar to the human fingernail. The almost finger-like dexterity available to cloven hoofed mammals like the Mountain Goat and Mountain Sheep combined with a hard outer shell and soft and flexible inner pads provide excellent traction in their precarious habitats.

Evolution

It is speculated that during the Eocene period, hoofed marsh dwellers carried their body weight using mainly two of the middle toes, which grew to be of equal size, becoming the Artiodactyla or even-toed hoofed animals. Prior to the close of the Eocene period the side toes of some had dwindled and practically disappeared while the basal pieces or metapodium of the pair of supporting toes became fused together, thus producing the appearance of a cloven hoof.

The mammal with a cloven hoof is an even-toed ungulate of order Artiodactyla as opposed to the odd-toed ungulates of Perissidactyla, like the horse, which have one toe, or the rhinoceros, which has three toes. The five-toed ancestors of the earliest Eocene had already developed feet that suggest odd-toed and even-toed descendants to the modern viewer. Even Phenacodus, the most generalized of the early mammals, has a foot in which the central toe is somewhat larger than the others and could be placed in the division of odd toed ungulates, Perissidactyla.

Cloven Hooves in Culture

Unclean Animals

In Judaism, Kashrut (*kosher*) is the set of dietary laws governing what may or may not be consumed. These laws are based upon the Torah and the Talmud. According to Jewish law, animals that *both* chew their cud (ruminate) and have cloven hooves are kosher . Animals with one characteristic but not the other (the camel, the hyrax and the hare because they have no cloven hooves, and the pig because it does not ruminate) are specifically excluded . Jews are forbidden from eating pork, as swine are considered to be unclean. In the context of Judaism, the books of Leviticus and Deuteronomy from the Hebrew Bible specify what must not be consumed. These are the animals considered to be unclean according to Leviticus 11 and Deuteronomy 14. Verse 7 to verse 8: Nevertheless these ye shall not eat of them that chew the cud, or of them that divide the cloven hoof; as the camel, and the hare, and the coney: for they chew the cud, but divide not the hoof; therefore they are unclean to you. And swine, because it divideth the hoof, yet cheweth not the cud, it is unclean unto you: ye shall not eat of their flesh, nor touch their dead carcase.

The Devil

The cloven hoof is also traditionally associated with the Devil. In works from Shakspeare and Sir Thomas Browne the association is very close and the devil takes pains to hide or disguise the hoof. In other works the Devil and the cloven hoof can be separated, some believe the cloven hoof more mischievous than the devil. But not every cloven hoof is the devil's minion. In Teutonic Mythology the devil has a non-cloven hoof that is solid like that of a horse, from which he can not be separated.

There is a story that in the year 1165 the Devil was seen riding like a great black horse before a storm in Yorkshire, the marks of his feet were visible in several places, particularly on the cliff at Scarborough, where he sprang into the sea. In the late 1800s, in the

South of England, was the discovery of marks in various parts of the country which could not be identified as any known beast or bird. The people did not like to say it was the Devil, but no explanation of the mystery was offered. Perhaps this extraordinary presence may have been nothing more than the cloven hoof which in the deep snows of winter is said to haunt the Dewerstone a rocky elevation on the borders of Dartmoor But this latter phenomenon is reported to be accompanied by a naked human foot. A case occurred in Devonshire and are many cases are collected in "Lancashire Folk Lore" 1867.

Eimer's Organ

Eimer's organs are sensory organs in which the epidermis is modified to form bulbous papillae. First isolated by Theodor Eimer from the European mole in 1871, these organs are present in many moles, and are particularly common in the star-nosed mole, which bears 30,000 of them on its unique tentacled snout. The organs are formed from a stack of epidermal cells, which is innervated by nerve processes from myelinated fibers in the dermis, which form terminal swellings just below the outer keratinized layer of epidermis. They contain a Merkel cell-neurite complex in the epidermis and a lamellated corpuscle in the dermal connective tissue.

Enteric Fermentation

Enteric fermentation occurs when methane (CH_4) is produced in the rumen as microbial fermentation takes place. Over 200 species of microorganisms are present in the rumen, although only about 10% of these play an important role in digestion. Most of the CH_4 byproduct is belched by the animal, however, a small percentage of CH_4 is also produced in the large intestine and passed out as gas.

In Australia ruminant animals account for over half of their green house gas contribution from methane. Australia has implemented a voluntary immunization program for cattle in order to help reduce flatulence-produced CH_4.

Although enteric fermentation is the largest anthropogenic source of methane emissions in the United States, in 2007, methane emissions from enteric fermentation were only 1.9% of total greenhouse gases produced in the United States in teragrams of carbon dioxide equivalents (Tg CO_2).

Epipubic Bone

Epipubic bones are a pair of bones which project forward from the pelvic bones of modern marsupials and of some fossil mammals-

multituberculates, monotremes, and even basal eutherians (the ancestors of placental mammals). In modern marsupials the epipubic bones are often called "marsupial bones" because they support the mother's pouch ("*marsupium*" is Latin for "pouch"). But their presence on other groups of mammals indicates that this was not their original function, which some researchers think was to assist locomotion by supporting some of the muscles that flex the thigh.

The epipubic bones were first described in 1698 but their functions have remained unresolved. It has been suggested that they form part of a kinetic linkage stretching from the femur on one side to the ribs on the opposite side. This linkage is formed by a series of muscles: each epipubic bone is connected to the femur by the pectineus muscle, and to the ribs and vertebrae by the pyramidalis, rectus abdominis, and external and internal obliques. The epipubic bones act as levers to stiffen the trunk.

Fur

Fur is a synonym for hair, used more commonly in reference to non-human animals, usually mammals; particularly those with extensive body hair coverage. The term is sometimes used to refer to the body hair of an animal as a complete coat, also known as the *"pelage"*. *Fur* is also used to refer to animal pelts which have been processed into leather with the hair still attached. The words *fur* or *furry* are also used, more casually, to refer to hair-like growths or formations; particularly when the subject being referred to exhibits a dense coat of fine, soft "hairs".

Animal *fur*, if layered, rather than grown as a single coat, may consist of short ground hair, long guard hair, and, in some cases, medium awn hair. Mammals with reduced amounts of fur are often called "naked", as in *The Naked Ape*, naked mole rat, and naked dogs.

The acquisition and use of fur as clothing and/or decoration is considered controversial in some cultures. Animal welfare advocates object to the trapping and killing of wildlife, and the confinement and killing of animals on fur farms. An animal with commercially valuable fur is known within the fur industry as a furbearer.

Nature of Fur

Fur usually consists of two main layers:

- *Ground hair or underfur or down hair* — the bottom layer consisting of wool hairs which tend to be shorter, flattened, curly, and denser than the top layer. Its principal function is thermal insulation and thus thermoregulation.

- *Guard hair* — the top layer consisting of longer, often coarser, straight shafts of hair that stick out through the underfur. This is usually the visible layer for most mammals and contains most of the pigmentation. It protects the underfur from outside factors, such as rain, and is thus often water-repellent..

A third layer, the awn hair, may also exist. It is intermediate between the two others.

Furstenberg's Rosette

Hock (Zoology)

The hock, or gambrel, is the tarsal joint of a digitigrade quadruped, such as a horse, cat, or dog. It is the anatomical homologue of the ankle of the human foot.

Horse

Although the *tarsus* refers specifically to the bones and joints of the hock, most people working with horses refer to the *hock* in such a way to include the bones, joints, and soft tissue of the area. The hock is especially important in equine anatomy, due to the great strain it receives when the horse is worked. Jumping, and movements that require collection, are some of the more stressful activities.

Primary Joints and Bones of the Hock

In the horse, the hock consists of multiple joints, namely:

- Tibiotarsal or Tarsocrural joint
- Proximal Intertarsal joint or Talocalcanealcentroquartal joint
- Distal Intertarsal joint or Centrodistal joint
- Tarsometatarsal joint
- Talocalcaneal joint.

In the horse, the hock consists of the following bones

- Talus
- Calcaneus
- Central tarsal bone
- 3rd tarsal bone
- 3rd metatarsal bone
- 4th tarsal bone
- Fused 1st and 2nd tarsal bone.

Equine Disease States

·Horses may suffer from "capped hock", which is caused by the creation of a false bursa, a synovial sac beneath the skin. Capped hock is usually caused by trauma such as kicking or slipping when attempting to stand. In the absence of a wound, it does not require immediate veterinary attention and is usually only of cosmetic significance. On the other hand, a wound into the calcanean bursa is a serious problem. A capped hock is extremely unlikely to be a cause of lameness, even if severe.

- Osteochondrosis dissecans, or OCD is a developmental defect in the cartilage or of cartilage and bone seen in particular locations on the surface of the tarsocrural joint. This condition is typically discovered when the horse is young, and is one cause of bog spavin. After surgery to remove bone and cartilage fragments most horses can return to full work.
- Distension of the tibiotarsal joint with excessive joint fluid and/or synovium is called bog spavin.
- Degenerative joint disease of the tarsometatarsal or distal intertarsal joint is referred to as bone spavin.
- Curb, or tarsal plantar desmitis, is traditionally considered a sprain of the plantar ligament, which runs down the back of the hock, serving functionally as a tension band connecting the calcaneus, the fourth tarsal bone and the fourth metatatarsal bone. Recent work has shown that curb can be caused by damage to one of many soft tissue structures in this region.
- Stringhalt.

Conformational Defects

Because the hock takes a great deal of strain in all performance disciplines, correct conformation is essential if the horse is to have a sound and productive working life. Common conformational defects include sickle hocks, post-legged conformation/straight hocks, cow hocks, and bowed hocks. Depending on the use of the horse, some defects may be more acceptable than others.

Hoof

A hoof is the tip of a toe of an ungulate mammal, strengthened by a thick horny (keratin) covering. The hoof consists of a hard or rubbery sole, and a hard wall formed by a thick nail rolled around the tip of the toe. The weight of the animal is normally borne by both the sole and the edge of the hoof wall.

Most even-toed ungulates (such as sheep, goats, deer, cattle, bison and pigs) have two main hooves on each foot, together called a cloven hoof. Most of these cloven-hoofed animals also have two smaller hoofs called *dew-claws* a little further up the leg – these are not normally used for walking, but in some species with larger dew-claws (such as deer and pigs) they may touch the ground when running or jumping, or if the ground is soft. Other cloven-hoofed animals (such as giraffes and pronghorns) have no dew claws. In some so-called "cloven-hoofed" animals such as camels, there are no hooves proper – the toe is softer, and the hoof itself is reduced to little more than a nail.

Some odd-toed ungulates (equids) have one hoof on each foot; others (including rhinoceroses, tapirs and many extinct species) have (or had) three hoofed or heavily nailed toes, or one hoof and two dew-claws. The tapir is a special case, with three toes on each hind foot and four toes on each front foot.

Hunter-Schreger Band

Hunter Schreger bands, commonly abbreviated as HSB, are features of the enamel of the teeth in mammals, mostly placentals. In HSB, enamel prisms are arranged in layers of varying thickness at about right angles to each other. HSB strengthen the enamel and prevent cracks from propagating through the tooth.

HSB are first observed in early Paleocene mammals, but at this time the HSB occupy only a small portion of the incisor and the angle between the bands is low. By the late Paleocene, HSB is seen to extend throughout the enamel and the bands are located at nearly right angles to each other.

Among Glires, the group containing rodents, lagomorphs, and their primitive relatives, the absence of HSB from the incisors has been considered primitive. Some early representatives, including *Eurymylus*, lack HSB, but others, including *Matutinia* and some mimotonids, have double-layered incisor enamel with HSB in the inner portion (portio interna, PI). Other mimotonids have single-layered enamel with HSB. All leporids studied also exhibit this pattern, except for an early Eocene, indeterminate leporid with HSB only in the PI. Ochotonids also have HSB in the PI only.

In rodents, HSB are usually present in the PI. Three types of HSB can be separated in rodents—pauciserial, uniserial, and multiserial. Pauciserial HSB, present in some primitive Paleogene rodents, are usually three to six prisms thick and have inter-prismatic matrix

(IPM) surrounding the prisms, which have irregular cross-sections. Uniserial HSB, present in most living rodents, consist of a single layer of prisms. Multiserial HSB, three to seven prisms thick, characterize the living Hystricognathi and the gundis (Ctenodactylidae) and springhares (*Pedetes*).

There are three subtypes on the basis of the orientation of the IPM. In the first, the IPM runs mostly parallel to the HSB, but does not surround it as in pauciserial enamel; in the second, the IPM makes an angle of about 45° with the HSB; and in the third, the two are located at right angles to each other. The last type serves to further strengthen the enamel and is characteristic of the Octodontoidea.

Among carnivorans, the hyenas exhibit specialized HSB, which form a complex, three-dimensional zigzag pattern, a specialization for bone-eating.

Interdigital Webbing

Interdigital webbing is the presence of membranes of skin between the digits. Normally in mammals, webbing is present in the embryo but resorbed later in development, but in various mammal species it occasionally persists in adulthood. In humans, it can be found in those suffering from LEOPARD syndrome and from Aarskog-Scott syndrome.

Webbing between the digits of the hindfoot is also present in several mammals that spend part of their time in the water. Webbing probably accommodates movement in the water.

Interdigital webbing is not to be confused with syndactyly, which is a fusing of digits and occurs rarely in humans. Syndactyly specifically affecting feet occurs in birds (such as ducks), amphibians (such as frogs), and mammals (such as the kangaroo).

Mammals with Interdigital Webbing

Rodents

In oryzomyines, a mainly South American rodent group, the Marsh Rice Rat, *Pseudoryzomys simplex*, and *Sigmodontomys alfari* all have small webs, which do not extend to the end of the proximal phalanges, whereas *Amphinectomys savamis*, *Lundomys molitor* and the members of the genera *Holochilus* and *Nectomys* have more expansive webbing, which extends beyond the proximal phalanges. Webbing apparently developed several times in oryzomyines and may also have been lost in some groups. Most ichthyomyines, an exclusively semiaquatic South and Central American rodent group, have small webbing, but members of the genus *Rheomys* have more expansive webs. Webbing is also present in the Australasian semiaquatic hydromyines (subfamily Murinae) of the genera *Baiyankamys*, *Hydromys*, and *Crossomys*; in the latter, it is most well-developed. The African semiaquatic rodents *Colomys goslingi* and *Nilopegamys plumbeus*, also members of the Murinae, lack interdigital webbing. Webbing is present in the hind feet of the Coypu (*Myocastor coypus*) of South America, which is currently classified in its own family.

Soricomorphs

Among shrews, the members of the genera *Chimarrogale* of southeastern Asia and *Neomys* of western Eurasia have interdigital webbing, as does the American Water Shrew (*Sorex palustris*) of North America, but it is more well-developed in *Nectogale elegans* of montane Asia. Webbing is also present in the Pyrenean Desman (*Galemys pyrenaicus*).

Tenrecs

The tenrec family, which occurs in Africa and mainly on Madagascar, includes several semiaquatic forms, and the small otter-shrews (*Micropotamogale*) and the aptly named Web-footed Tenrec (*Limnogale mergulus*) have developed interdigital webbing.

Opossums

The Water Opossum (*Chironectes minimus*) of South America is the only opossum with interdigital webbing.

Carnivorans

Several carnivorans have interdigital webbing, including the Greater Grison (*Galictis vittata*), the Colombian Weasel (*Mustela felipei*), the Amazon Weasel (*Mustela africana*), and the Mink (*Neovison vison*).

All otters have interdigital webbing, in the fore or hind limbs or both, to aid in aquatic propulsion. In sea otters, the webbing is covered with hair, at a density of 3300 hairs per square centimeter.

Whales

Pits present on the sides of fossil proximal phalanges of pakicetids, ancestral whales, suggest that these animals had interdigital webbing, a development hypothesized to lead to the fluke, spurred by FGF8, a Fibroblast growth factor.

Malleus

The malleus is unique to mammals, and evolved from a lower jaw bone in basal amniotes called the articular, which still forms part of the jaw joint in reptiles. Embryologically it is derived from the first pharyngeal arch along with the rest of the bones of mastication, such as the maxilla and mandible.

Omasum

The omasum, also known as the bible , the fardel , the manyplies and the psalterium , is the third compartment of the stomach in ruminants. Though its functions have not been well-studied, it appears to primarily aid in the absorption of water, magnesium, and the volatile fatty acids produced by rumen fermentation, that have not been absorbed into the bloodstream yet. The numerous folds of its mucosa are thought to trap digesta particles so that the maximum amount of nutrients may be absorbed.

There is some evidence that the contractions of the omasum can propel large particles backwards through the reticulorumen orifice, the junction connecting the omasum with the reticulorumen, into the reticulorumen, the first compartment of the ruminant stomach. In this way, the omasum allows large particles, which still likely contain appreciable amounts of fermentable substrate, to be further digested in the reticulorumen. (Though fermentation initiated in the reticulorumen can continue in the omasum, it does so in only limited quantities, so this mechanism of ejecting largely unfermented particles into the reticulorumen is necessary for complete fermentation.)

Paw

A paw is the soft foot of a mammal, generally a quadruped, that has claws or nails. A hard foot is called a hoof. Paws are used to pad feet for walking and increase friction.

Common Characteristics

The paw is characterised by thin, pigmented, keratinised, hairless epidermis covering subcutaneous, collagenous, and adipose tissue, which make up the pads. These pads act as a cushion for the load-bearing limbs of the animal. The paw consists of the large, heart-shaped metacarpal pad (forelimb) or metatarsal pad (rear limb), and generally four load-bearing digital pads, although there can be five or six toes in the case of bears and the Giant Panda. A carpal pad is also found on the forelimb which is used for additional traction when stopping or descending a slope in digitigrade species. Additional dew claws can also be present. The paw also includes a horny, beak shaped claw on each digit. Though usually hairless, certain animals do have fur on the soles of their paws.

An example being the Red Panda, whose furry soles help insulate them in their snowy habitat.

Rump (Animal)

The rump or croup, in the external morphology of an animal, is the portion of the posterior dorsum that is posterior to the loins and anterior to the tail. Anatomically, the rump corresponds to the sacrum.

The tailhead or dock is the beginning of the tail, where the tail joins the rump. It is known also as the base or root of the tail, and corresponds to the human sacrococcygeal symphysis.

In certain mammals the tail may be said to consist of the tailbone (meaning the bony column, muscles, and skin) and the skirt (meaning the long hairs growing from the tailbone). In birds, similarly, the tail consists of tailbone and tailfan (tail fan).

Some animals are subjected to docking, the amputation of the tailbone at or near the dock. These include dogs, cats, sheep, pigs, and horses. Humans have a remnant tail, the coccyx, and the human equivalent of docking is coccygectomy.

Usage

Usage varies from animal to animal. Birds and cattle are said to have a rump and tailhead. Dogs are said to have a rump and dock. Horses are said to have a croup (sometimes rump) , thigh or haunch, buttock, and dock.

Birds

In bird anatomy, the rump is the body immediately above the tail. The colour of plumage on the rump is a characteristic widely used by ornithologists to distinguish between related species, and sometimes also between males and females of the same species.

Similarly, the silhouette of the tailfan is a characteristic widely used for purposes of identification, particularly in the field.

Some birds have a food reservoir pouch in the esophagus that is known as a croup.

Dogs

In some breeds it is traditional for tails to be cut off (docked) at the dock.

Horses

In horse anatomy, the croup refers specifically to the topline of the horse's hindquarters and surrounding musculature, beginning at the hip, extending proximate to the sacral vertebrae and stopping at the dock of the tail (where the coccygeal vertebrae begin).

Below the croup is the Thigh or Haunch. Behind the thigh is the buttock. On horses appearing in parades and other public ceremonies, the croup may be decorated with a pattern in the horse's hair, formed by applying hair gel or spray, then brushing patches of hair in opposite directions.

Applied to horses, the term dock has two additional uses. Its meaning may be extended to the entire tail minus the skirt (i.e., synonymous with tailbone) or to the tailhead only. In other equidae, it encompasses most of the tailbone, as most of that portion of the tail does not have long hairs. A lack of long hairs can be natural, as in zebras, donkeys, and the Przewalski horse, or artificial, the result of pulling, trimming, or shaving part of the skirt.

A sponge used to wash the hairless skin on the underside of the dock and other regions under the tail, protected by the dock, is called a dock sponge. Thus, the meaning of "dock" has been used to refer to the orifices beneath the dock, specifically the anus and vagina, creating a misapprehension that "dock" refers to the anus, as in, *a horse's fundamental orifice is its dock.*

Stringhalt

Stringhalt Is a sudden flexion of one or both hind legs in the horse. Most easily seen while horse is walking, or trotting. Best while backing up slowly, turning on affected leg, or suddenly frightened. It involves one, or both hind limbs of the horse. It is a spasmatic contraction of the lateral extensor tendons of the hind legs.

There are 4 forms of stringhalt:

1. Australian stringhalt: a false dandelion (catsear) is the suspected cause
2. Pseudostringhalt: the apparent result of a painful condition in the affected leg
3. Unilateral stringhalt: cause unknown, but often relieved by section of the lateral extensor tendon
4. Bilateral stringhalt: in two out of two affected horses necropsied, abnormal thalamus.

Treatment Varies

Teat

Teat is an alternative word for the nipple of a mammary gland, in humans referred to as a breast, from which milk is discharged. Similarly, in cows, goats, etc., teats are the projections from the udder through which milk is discharged.

The rubber mouthpiece of a baby bottle and pacifier may also be referred to as a teat.

The word "teat" is of Germanic origin, having the same origin as the Dutch word "tiet" and German "Zitze". In turn, this word has Indo-European roots, as may be seen in the Welsh word "teth". The Old English for teat was "tit", which is still used as a vulgar or slang term for "breast" in Modern English.

The number of teats varies in mammals from 2 to 20. Marsupials and eutherian mammals have teats from which their young suckle milk. monotremes lack teats; their young drink milk directly from pores in the skin (similar to sweat glands), or by sucking it off of hairs surrounding the pores. In most eutherian mammals, both males and females have teats. Those of the male are nonfunctional except in cases of hormonal imbalance. Notable exceptions to this are rats and horses, neither of which has teats in the male.

Teats vary in size, location, and structure in different mammalian species. Teats, on occasion, become so plump and filled that milk may escape without suckling. Female goats and ewes have two teats, each with a single mammary gland, located between the hindlegs. Mares have two teats, each with two mammary glands. The teats of the sow can be quite variable in number, from six to thirty, and are located on two parallel lines along the belly. Cows have four teats, each with one mammary gland in the udder. Humans have two teats. Extra teats occur often, and are known as supernumerary teats. They are nonfunctional.

The offspring of domestic animals, including piglets, calves, lambs, and foals, engage in a behaviour known as teat seeking, sipping, and suckling. This strong instinct occurs in most species within minutes of birth, and serves both to connect the young to the food source and to encourage bonding between mother and young. The offspring thoroughly enjoy their suckling time, and may suckle even after filling up. Some teats may become swollen and tender after a large amount of suckling.

Udder

An udder is an organ formed of the mammary glands of female quadruped mammals, especially ruminants such as cattle, goats, sheep and deer. The udder is a single mass hanging beneath the animal, consisting of pairs of mammary glands. In cattle there are normally two pairs, in sheep, goats and deer there is one pair, and in some animals such as pigs there are many pairs. Udder care and hygiene in cows is important in milking, aiding uninterrupted and untainted milk production, and preventing mastitis. Products exist to soothe the chapped skin of the udder.

Ungual

An ungual (from Latin *unguis*, i.e. *nail*) is a highly modified distal toe bone which ends in a hoof, claw, or nail. Elephants and other ungulates have ungual phalanges, as did the sauropods and horned dinosaurs. A claw is a highly modified ungual phalange.

Whiskers

Vibrissae (singular: vibrissa), or whiskers, are specialized hairs (or, in certain bird species, specialized feathers) usually employed for tactile sensation. Vibrissae hairs commonly grow around the nostrils, above the lips, and on other parts of the face of most mammals, as well as on the forelegs and feet of some animals. Vibrissae are usually thicker and stiffer than other types of hair. The term may also refer to the thick hairs found inside human nostrils, but these have no sensorial function and only operate as an airborne particulate barrier.

Vibrissae consist of inert material and contain no nerves, but do have special sensory cells associated with them. Vibrissae are different from other hairs mainly because they are implanted in a special follicle sealed by a capsule of blood, called a blood sinus.

Touching a vibrissa causes it to bend, and the blood in the sinus is pushed to one side or the other. The blood amplifies the movement and allows the mechanoreceptors at the base of the vibrissa to detect extremely small deflections.

In some mammals, the follicles of vibrissae are surrounded by a highly developed sheath of muscle tissue which can be used to move them, such as in the case of whiskers found on cats, dogs and other mammals. Whiskers can grow to be extremely long; the length of a chinchilla's whiskers can be up to a third of its body length. Vibrissae offer an advantage to most animals that do not always have sight to rely on to navigate or to find food, or when the usefulness of non-tactile senses is limited. Some animals, such as house mice, can even detect air movements with their vibrissae.

A large part of the brain of many mammals is devoted to processing the nerve impulses from vibrissae, because this is important for survival. Information from the vibrissae is transmitted and processed through the trigeminal nerve into the brainstem and thalamus before relaying to the barrel cortex of the brain. Mammals use a great deal of energy to keep the follicles housing their whiskers warm and ready to use. Some animals-mainly rodents-actively palpate their vibrissae, a process known as whisking, whilst others use them merely as passive sensors. The arrangement of whiskers is not random: a good picture will show a grid of three lines on either side of the face, each line has 6 to 8 whiskers, shorter in front and longer at the rear.

Zygomatic Plate

In rodent anatomy, the zygomatic plate is a bony plate derived from the flattened front part of the zygomatic arch (cheekbone). At the back, it connects to the front (maxillary) root of the zygomatic arch, and at the top it is connected to the rest of the skull via the *antorbital bridge.* It is part of the maxillary bone, or upper jaw, which also contains the upper cheekteeth. Primitively, rodents have a nearly horizontal zygomatic plate. In association with specializations in zygomasseteric system, several distinct morphologies have developed across the order. The term is also used for an analogous structure in some South American typotheres, including *Pseudotypotherium* and *Medistylus.*

Function

The zygomatic plate serves to resist muscular tension resulting from the contraction of the incisors by the anterior deep masseter muscle; thus, rodents which pulverize hard food with the incisors tend to have broader zygomatic plates than those that rather use their molars for this purpose.

Hystricomorpha

The members of this large and diverse suborder have a narrow, low zygomatic plate.

Sciuromorpha

The suborder Sciuromorpha includes three families. Squirrels (family Sciuridae) tend to have broad zygomatic plate that extend above the infraorbital foramen. The mountain beaver (*Aplodontia rufa*), the only surviving member of its family, retains the primitive narrow and low plate. The dormice (Gliridae) have broad, high zygomatic plates, except for *Graphiurus*, which has a lower plate.

Castorimorpha

Members of the suborder Castorimorpha, which includes the beavers, pocket gophers, and pocket mice, tend to have broad zygomatic plate that extend above the infraorbital foramen.

Anomaluromorpha

Anomaluromorpha is a small suborder, containing only two families. Anomaluridae have a low and narrow zygomatic plate. Members of the subfamily Idiurinae are atypical in having the zygomatic plate extended forward nearly to the incisors. The condition in the springhaas (*Pedetes*) is similar.

Myomorpha

Myomorpha is the largest suborder of rodents. In the most numerous subgroup, the Muroidea (including all living families except Dipodidae), the zygomatic plate is generally broad and tilted upwards. Muroids may have the plate extending in front of the front (maxillary) root of the actual zygomatic arch, creating a *zygomatic notch.*

In some, the plate extends at the front into a spinous process, the *zygomatic spine.*

Dipodidae

Members of the family Dipodidae, which have hystricomorphous zygomasseteric morphology, have nearly horizontal, narrow zygomatic plates.

Platacanthomyidae

Members of the small family Platacanthomyidae have a relatively narrow zygomatic plate.

Spalacidae

The fossorially specialized family Spalacidae shows peculiarities in the condition of the zygomatic plate. In *Tachyoryctes* and the Rhizomyinae, it is tilted upward and fused to the sides of the snout (rostrum).

In the Spalacinae and Myospalacinae, on the other hand, the plate is tilted downward into an almost horizontal position.

Calomyscidae

The mouse-like hamster (*Calomyscus*), the only member of its family, has a straight front margin on the zygomatic plate.

Nesomyidae

Brachyuromys has an arvicoline-like high zygomatic plate. In *Eliurus*, the front border of the plate is straight. *Nesomys* has a low zygomatic plate. In *Hypogeomys*, it is broad, but rather low.

Muridae

Muridae is the order's largest family, and contains several subfamilies.

Deomys, a member of the Deomyinae, has an unusually low zygomatic plate, as does *Lophuromys*, a member of the same subfamily.

Most members of the subfamily Murinae, the Old World rats and mice, have a fairly broad zygomatic plate with a well-developed zygomatic notch. A zygomatic spine is developed in some Australian genera, including *Notomys* and some *Pseudomys*. Except for *Xeromys*, *Hydromys* and related genera ("hydromyines") have a narrow plate, lacking the notch, as does *Hyomys*, *Macruromys*, *Crunomys*, and *Rhynchomys*. The Philippine *Batomys*, *Carpomys*, and *Crateromys* have well-developed zygomatic plates, reminiscent of those in Arvicolinae. *Phloeomys* has a broad zygomatic plate.

Cricetidae

The family Cricetidae is the order's second largest, containing several subfamilies and hundreds of species.

The subfamily Arvicolinae, the voles and lemmings, has the zygomatic plate tilted upwards very strongly.

In the subfamily Tylomyinae, *Nyctomys* has a narrow zygomatic plate.

Among members of the Neotominae, *Baiomys*, *Reithrodontomys*, *Onychomys*, and *Peromyscus* has a narrow zygomatic plate.

Members of the subfamily Sigmodontinae, which includes a number of tribes, usually have the antorbital bridge below the upper surface of the skull. Most have a zygomatic notch.

The extent of the zygomatic plate at the back is also variable within Sigmodontinae, with some having the plate extending back to the level of the first upper molar and others having shorter plates. Members of the semiaquatic tribe Ichthyomyini are unique among the

Sigmodontinae in lacking the zygomatic notch. In ichthyomyines, the development of the zygomatic plate is correlated with the development of the teeth: those species with large molars and small incisors, including species of *Anotomys* and *Rheomys*, have slender plates that do not extend back to the first molars, whereas those with larger incisors and smaller molars, including some *Ichthyomys* and *Neusticomys*, have broader zygomatic plates that do reach the level of the first molars.

The genus *Sigmodon*, which is classified in its own tribe, has a broad zygomatic plate and a zygomatic spine. Relative width of the zygomatic plate can distinguish some species of *Sigmodon*. Most members of the tribe Phyllotini have the antorbital bridge located higher than is usual in Sigmodontinae (*Calomys* and *Andalgalomys* show the normal sigmodontine condition). A similar condition characterizes *Euneomys*, *Neotomys*, *Reithrodon*, which are no longer considered phyllotines, but to an even larger extent than in most actual phyllotines; in *Euneomys*, the antorbital bridge is inserted on the upper surface of the skull. Most phyllotines have zygomatic spines, but the structure is more well-developed in *Reithrodon*. The zygomatic plate not extending backwards to the first molars is a diagnostic character of phyllotines.

Most thomasomyines lack a well-developed zygomatic notch. The genus *Rhipidomys* has a narrow zygomatic plate, no zygomatic spine and only a narrow notch. *Thomasomys* shares a narrow zygomatic plate. In the Akodontini, *Oxymycterus* and *Lenoxus* have a low zygomatic plate, similar to that of *Lophuromys*. *Scapteromys* shares a low plate. In *Akodon*, the plate is narrow, as in many others akodontine; this is extremely so in *Blarinomys*. The tribe Abrotrichini is characterized by a narrow zygomatic plate, without an extension at the upper border.

In the tribe Oryzomyini, the configuration of the zygomatic plate is variable. Most have a well-developed zygomatic notch. In the three related genera *Holochilus*, *Pseudoryzomys*, and *Lundomys*, this extension has further developed into a zygomatic spine. In contrast, *Microryzomys*, *Oreoryzomys*, *Oecomys*, *Scolomys*, and *Sigmodontomys aphrastus* lack a well-defined notch and do not have the plate extending appreciably in front of the root of the zygomatic arch. The zygomatic plate extending back to the level of the upper first molar is a putative synapomorphy of Clade C within Oryzomyini. The sigmodontine *Juliomys pictipes* has an almost vertical zygomatic plate.

7

The Tissues and Organs of the Lymphatic System

The lymphatic system is a specialized form of connective tissue. It consists of groups of cells, tissues and organs that monitor body surfaces and internal fluid compartments and react to the presence of potentially harmful antigenic substances. These substances include infectious micro-organisms, toxins, foreign cells and tissues, as well as normal cells that have transformed into cancer cells and are recognized as "non-self". Lymphatic tissue constitutes the second line of defence against invaders. The first line of defence is the epithelial covering of the skin, and the gastrointestinal, respiratory and urogenital tracts. Lymphocytes are the chief cellular constituents of lymphatic tissue and are the key element in an immune response. Functionally, there are two types of lymphocytes, B lymphocytes and T lymphocytes. Both types differentiate from precursors called colony-forming units (CFUs) that originate in early fetal life in the yolk sac, liver and spleen. By late fetal life, CFUs are restricted to the bone marrow, where they continue to proliferate during postnatal life.

Differentiation of Lymphatic Tissue

Primary Lymphatic Organs and Tissues

Undifferentiated stem lymphocytes migrate to the primary lymphatic organs and tissues where they undergo differentiation into immunocompetent cells. This process entails the synthesis of unique internal and membrane proteins. This initial proliferation and differentiation is antigen-independent, and gives rise to populations of cells in which individual cells are genetically pre-programmed to recognize only one antigenic determinant. B and T lymphocytes undergo differentiation in different areas. T lymphocytes or T cells undergo differentiation in the thymus (hence "T"). They have a long life span and are involved in cell-mediated immunity. B lymphocytes or B cells undergo differentiation in special microenvironments in the bone marrow. Recent evidence suggests that the lymphatic tissue of the cecum, appendix and ileum is also involved.

These areas of the bone marrow and gut-associated lymphatic tissue (GALT) are sometimes called the "bursa equivalent" areas, after the bursa of Fabricius, a mass of lymphoid tissue associated with the cloaca, in birds. B cells got their name because B cell differentiation was first demonstrated in the bursa of Fabricius of chicken embryos. B cells are involved in humoral immunity and the production of antibodies.

To summarize: The primary lymphatic organs are the thymus (for T cells) and bone marrow and probably specific areas of GALT, collectively called bursa equivalent, (for B cells). The primary lymphatic organs are sometimes called the central lymphatic organs.

Secondary Lymphatic Organs and Tissues

Immunocompetent lymphocytes enter the blood and lymph systems to be transported throughout the body. Together with plasma cells, derived from B lymphocytes, and macrophages, they organize around mesenchymal reticular cells and their reticular fibres to form the secondary lymphatic organs (also called the peripheral lymphatic organs).

The secondary lymphatic organs and tissues are lymph nodules, the lymph nodes, tonsils and spleen. Immunocompetent lymphocytes also travel to the diffuse connective tissue (lamina propria) that underlies the epithelium of the digestive, respiratory and urogenital tracts.

Histology of the Lymphatic Organs and Tissues

Diffuse Lymphatic Tissue

The diffuse lymphatic tissue underlying the epithelia of various organ systems is important in monitoring against invaders. In the event that a lymphocyte reacts with an antigen, it travels to a regional lymph node and undergoes proliferation and differentiation. Its progeny then return to the lamina propria as effector B or T lymphocytes, plasma cells and memory cells. The regular presence of large numbers of plasma cells (especially in the lamina propria of the GI tract) indicates local antibody secretion.

Lymph Nodules

Sometimes lymphocytes form localized concentrations that are sharply defined but not encapsulated. Such concentrations are called lymph nodules. They are common in the walls of the alimentary canal, respiratory passages and urogenital tract, as well as other parts of the body. For the most part, such nodules are disposed singly and in a random manner. However, in the alimentary canal aggregations of nodules are found in specific regions, notably the appendix and cecum and in the Peyer's patches of the ileum. A nodule consisting mainly of small lymphocytes is called a primary nodule.

A primary nodule appears uniformly dense or colored. Many nodules however, are secondary nodules, containing a lighter-staining central region called the germinal centre, and an outer ring of small lymphocytes. The germinal centre develops when a lymphocyte that has recognized an antigen returns to a primary nodule and undergoes blastic transformation. It is lighter-staining due to the presence of large lymphocytes (lymphoblasts and plasmablasts). Frequently.

Lymph Nodes

Lymph nodes are small, encapsulated lymphatic organs located along lymphatic vessels. They are bean-shaped, with a convex and a concave surface. They range in size from 1 mm to about 2 cm along their longest dimension. They serve as filters through which lymph percolates on its way to the blood. Although they are widely distributed, they are concentrated in certain areas, such as the groin, axilla, and mesenteries.

Supporting Framework of Lymph Node

The lymph node is covered by a capsule of dense connective tissue. Trabeculae, also made of dense CT, extend from the capsule into the substance, forming the gross framework of the node. A fine supporting meshwork of reticular fibres (made of collagen type III) is found throughout the remainder of the gland. Reticular fibres are not seen in routine preparations, but can be seen with special stains, eg. silver stains. The reticular fibres are secreted by reticular cells, which are stellate or elongate cells with an oval nucleus and a small amount of acidophilic cytoplasm. Cytoplasmic processes of reticular cells wrap around the bundles of reticular fibres, isolating them from the lymphocytic parenchyma. Although they can't be distinguished with the light microscope, two populations of reticular cells have been identified by electronmicroscopy, immunocytochemistry and autoradiography. The first population, as described above, has a structural function. The second population of reticular cells, also called dendritic cells, has functions characteristic of antigen-presenting cells or tissue macorphages. These cells are splayed along the meshwork of reticular fibres to increase the surface area for phagocytosis and other cell interactions.

Organization of Parenchyma of Lymph Node

The parenchyma of the lymph node is separated from the capsule by a sinus, the subcapsular (or marginal or cortical) sinus. The subcapsular sinus is continuous with sinuses that extend along the trabeculae, called trabecular sinuses.

The parenchyma of the lymph node is divided into a cortex and a medulla. The cortex forms the outer part of the node, except at the hilum. It consists of a dense mass of lymphatic tissue. In the outer part of the cortex, the lymphocytes are typically organized into nodules which consist mainly of B cells. The part of the cortex adjacent to the medulla is not organized into nodules. It is called the deep cortex (or juxtamedullary cortex, or paracortex). It is dependent on T cells for its development, and perinatal thymectomy results in a poorly developed paracortex. For this reason, it is also called the thymus-dependent cortex. The medulla forms the inner part of the lymph node. It consists of cords of lymphatic tissue, separated by lymphatic sinuses called medullary sinuses. The medullary sinuses are continuous with the trabecular sinuses of the cortex. They converge toward the hilum, where they drain into efferent lymphatic vessels. The medullary cords consist mostly of B lymphocytes.

Lymphatic and Blood Supply of Lymph Node

Afferent lymphatic vessels penetrate the capsule at various points on its convex surface. Lymph flows into the subcapsular sinus and through the trabecular and medullary sinuses toward the efferent lymphatic vessels. Efferent lymphatic vessels leave at the hilum, a depression on the concave surface, which also serves as the entry and exit for blood vessels and nerves. The spaces of the sinuses are crossed by a network of reticular fibres, reticular cells, and macrophages. Potential antigens, such as macromolecules, bacteria, parasites, or tumoral cells reaching the node through the afferent lymphatics are usually trapped in this network and processed by macrophages.

The sinuses have a lining of endothelium that is continuous when it is directly adjacent to the connective tissue of the capsule or trabeculae (septa), but discontinuous when it faces the lymphatic parenchyma. Lymphocytes and macrophages readily pass back and forth between the sinuses and parenchyma. A macrophage in the parenchyma may also send long cytoplasmic processes, called pseudopods, into a sinus through endothelial discontinuities to monitor the lymph percolating through. Arteries enter at the hilum and progress toward the cortex, forming capillary beds around the cortical nodules. Postcapillary venules progress back toward the hilum, becoming larger and converging onto the vein as they do so. Postcapillary venules are lined by an unusual cuboidal or columnar epithelium.

This extra-tall epithelium allows lymphocytes to pass through the vessel by diapedesis but prevents the passage of fluid. In fact, most lymphocytes enter the node through the postcapillary venules, although some enter with the lymph of the afferent lymph vessels. Antigen-transformed lymphocytes remain in the lymph node to proliferate and differentiate. T cells remain in the thymus-dependent paracortex, while B cells migrate to the nodular cortex. Most lymphocytes leave the lymph node by entering a lymphatic sinus from which they reach the efferent lymphatic vessels.

Tonsils

Tonsils are organs composed of aggregates of incompletely encapsulated lymphoid tissue. They are characterized by depressions of surface epithelium around which aggregations of lymph nodules are grouped. Three consular groups, the palatine tonsils, lingual tonsils, and pharyngeal tonsils, form a ring of lymphoid tissue surrounding the pharynx where the nasal and oral passages unite.

Tonsils aid in the protection of the body against invading bacteria, viruses, and foreign proteins. The antigens stimulate the production of antibodies in plasma cells derived from lymphocytes. However, epithelial erosion seems to enhance an invasion by microorganisms, and the tonsils are frequent portals of infection. Tonsils reach their maximum development in childhood, and thereafter form an incomplete ring around the pharynx.

The palatine (or faucial) tonsils are paired, ovoid structures located in the mucous membrane at the junction of the oropharynx and oral cavity (faucia). The lingual tonsils are small and fairly numerous structures located in the root of the tongue, behind the circumvallate papilla. The pharyngeal tonsil is a single tonsil located in the median posterior wall of the nasopharynx. (Extensions of the pharyngeal tonsil around the pharyngeal orifices of the eustacean tubes are sometimes considered as separate tonsils, the tubal tonsils.) Here, we will only look at the palatine tonsils.

The Palatine Tonsils

The palatine tonsils are covered on their free surface with stratified squamous epithelilum that is continuous with the lining of the mouth and pharynx. The epithelium rests on a basement membrane, under which there is a thin layer of fibrous connective tissue.

In about 10 to 20 places, the epithelium covering the tonsil dips into the interior of the tonsil, forming tonsillar crypts. The crypts are lined with a continuation of the stratified squamous epithelium. The lumens of the crypts contain desquamated epithelial cells, live and dead lymphocytes, and bacteria.

Lymphoid tissue surrounds the crypts as a diffuse mass in which lymph nodules are embedded. The nodules may contain germinal centres. In the deeper parts of the crypts, there is an intense infiltration of the epithelium with lymphocytes, and consequently no clear delineation between epithelium and lymphoid tissue. Adjacent to the deepest portions of the tonsil, the fibrous tissue is condensed to from a thin capsule that covers the base and sides of the tonsil.

The capsule acts as a barrier against spreading tonsillar infections. Connective tissue septa extend into the interior of the tonsil and separate the various crypts, with their surrounding lymphatic tissue, from one another.

Small mucous glands lie in the connective tissue beneath the tonsil and its capsule; their ducts usually open on the free surface, occasionally into the crypts.

Tonsils possess no afferent lymphatic vessels. Lymphoid tissue is supplied from blood vessels coursing in the capsule and septa of the tonsils. Plexuses of lymph capillaries occur around the lymphoid tissue and drain into efferent lymphatic vessels.

The Spleen

The spleen is the largest of the lymphoid organs. It is located in the upper left of the abdominal cavity and has a rich blood supply. While lymph nodes serve as the immunological filter of the lymph, the spleen serves as the immunological filter of the blood. It is an important defence against microorganisms that penetrate the circulation.

General Structure

The spleen is surrounded by a capsule, which itself is covered by a serous membrane, the peritoneum. Many trabeculae pass from the capsule into the interior of the organ. On the medial surface of the spleen is a deep indentation, the hilum, which provides for the passage of the splenic artery and vein, lymph vessels and nerves. The capsule and trabeculae are made of dense connective tissue that contains myofibroblasts. These cells are contractile and also produce the extracellular CT fibres. In those species in which the spleen has the capacity to hold a large volume of blood (not humans), contraction of the capsule and trabeculae helps discharge stored RBCs into the circulation.

The trabeculae, which branch and anastomose repeatedly, delineate many compartments, or lobules, within the spleen. Each lobule is bounded by several trabeculae, but the lobules are not distinct because they are not completely outlined by trabeculae. Each lobule is supplied by a central artery (described below) and drained by veins that run in the trabeculae to leave the lobule.

The parenchyma of the spleen is supported by a fine meshwork of reticular fibres that blends with the capsule, trabeculae and walls of blood vessels. Reticular cells (as described for lymph nodes) and macrophages are associated with the reticular fibres.

White Pulp and Red Pulp

The substance of a freshly cut spleen appears as two colours. Circular or elongated whitish-gray areas are surrounded by more reddish areas. These areas are called the white pulp and the red pulp, respectively. The white pulp consists of aggregations of lymphocytes, the red pulp consists of cords of cells separated by sinusoids.

Branches of the splenic artery course through the capsule and trabeculae and enter the parenchyma. The arteries, whose adventitia is largely replaced by reticular tissue, immediately become enveloped by a sheath of lymphocytes. This lymphocyte sheath is called the periarterial lymphatic sheath (PALS). It constitutes the white pulp, and the artery inside is called the central artery. In cross section, the sheath of lymphocytes appears circular and may give the appearance of a nodule; however, the presence of a central artery distinguishes PALS from a nodule. Nodules do, however, appear along the length of the periarterial lymphatic sheath. They appear as localized expansions of PALS and tend to displace the central artery. In cross sections of PALS with a nodule, the central artery has an eccentric, rather than a central position. The central artery sends branches into the PALS and to the sinus surrounding the PALS. The nodules of the white pulp consist of B lymphocytes, while the non-nodular PALS ensheathing the central artery consists of T lymphocytes (and is therefore a thymus-dependent zone). The nodules often have a germinal centre, which, as in other lymphatic tissue, is a response to antigen exposure. The periarterial lymphatic sheath is surrounded by a margin of less densely packed lymphoid tissue with many sinuses called the marginal zone. The marginal zone has fewer lymphocytes but many macrophages that show active phagocytosis. This zone plays a major role in the immunologic activity of the spleen. The perimeter of the white pulp is bounded by a sinus called the marginal sinus. Don't expect to be able to distinguish the marginal sinus on your slides.

The red pulp consists of cellular cords, called splenic cords or Billroth's cords, which form a spongy network of modified lymphatic tissue that gradually merges into the white pulp. Separating the cords are venous sinuses. The splenic cords are supported by the typical meshwork of reticular cells and fibres that contains large numbers of RBCs, macrophages, lymphocytes, plasma cells and granulocytes. The macrophages phagocytose (among other things) damaged RBCs. They begin the process of hemoglobin breakdown and iron reclamation. The iron is stored as ferritin or hemosiderin to be reutilized in the formation of new RBCs. It will be transported, in combination with transferrin, to the bone marrow to be reused in erythropoiesis.

The heme is broken down into bilirubin, to be transported to the liver via the portal system and there conjugated to glucuronic acid, which is secreted into the bile. Globin, the protein part of hemoglobin, is hydrolized to amino acids that are reutilized for protein synthesis. Of all the phagocytes in the body, the splenic ones are also the most active in the phagocytosis of living particles such as bacteria and viruses. The endothelial cells that line the sinuses of the red pulp are extremely long, and their long axes parallel the direction of the vessel. Electron microscopy has revealed that these rod-like endothelial cells are only intermittantly connected to one another by side processes and that they do not have a continuous basement membrane. Rather, strands of basement membrane loop around the outside of the sinus at right angles to the long axes of the endothelial cells, like hoops around the staves of a barrel. Thus, there are prominent intercellular spaces between the endothelial cells. Processes of macrophages extend between the endothelial cells and into the lumen of the sinuses to monitor the passing blood for antigens. Blood is also able to pass into and out of the sinuses quite easily. Blood fills both the sinuses and the cords of the red pulp, often to such an extent that it is difficult, in (light microscopy) histological sections, to distinguish between the cords and the sinuses.

Splenic Circulation-open and Closed Models

After numerous divisions, the arterioles of the PALS lose their investment of white pulp and enter the red pulp. Here each arteriole divides into several small branches that lie close together, like a brush or penicillus. These arterioles are called penicilli or penicillar arterioles. The penecilli continue as arterial capillaries. They may become surrounded by aggregations of macrophages, in which case they are called sheathed capillaries. The macrophages ingest blood-borne particles.

The manner in which blood flows from the arterial capillaries in the red pulp into the interior of the sinuses has not been completely elucidated. Some investigators believe that the capillaries open directly into the sinusoids; they are proponents of the closed circulation model. Other investigators believe that the blood empties directly into the reticular meshwork of Billroth's cords. The blood would then percolate throught the cords and be exposed to their macorphages before returning to the circulaton by entering a sinus from the extravascular side. This is the open circulation model.

The open circulation model is currently the most favored. It provides a more efficient exposure of blood to red pulp macrophages. In the closed circulation model, blood cells would have to leave the sinuses and then reenter them in order to be exposed to macrophages. All investigators agree that blood cells do leave the vasculature to populate the red pulp and white pulp and that they reenter the vascular system in the red pulp. (The sinuses, as described above, are very leaky.)

The venous sinuses empty into the pulp veins that leave the pulp and unite to form larger veins which pass into the trabeculae as trabecular veins. The trabecular veins (consisting only of endothelium and the fibromuscular tissue of the trabeculae) travel to the hilum, where they drain into the splenic vein.

Lymphatic vessels originate in the white pulp near the trabeculae and constitute a route for lymphocytes to leave the spleen. Efferent vessels are present in the larger trabeculae and capsule. The spleen has no afferent lymphatic vessels.

Functions of the Spleen

Its immune functions include:

- proliferation of lymphocytes
- production of humoral antibodies
- removal of macromolecular antigens from the blood

Its hematopoetic functions include:

- formation of blood cells in fetal life
- removal and destruction of senile, damaged and abnormal RBCs and platelets
- retrieval of the iron from hemoglobin
- storage of blood in some species (not humans).

Despite its many functions, the spleen can be removed surgically if necessary, for example, if trauma causes intractable bleeding from the spleen. The destruction of RBCs then occurs in the bone marrow and liver.

The Thymus

The thymus is a 2-lobed organ located in the upper mediastinum anterior to the heart and great vessels. The two lobes are united by connective tissue. The thymus varies in size and development with the age of the individual. It attains its maximum development after puberty, after which it becomes inconspicuous. Much of the thymus tissue eventually becomes replaced with adipose tissue.

The thymus is the site where stem lymphocytes proliferate and differentiate into T cells. During fetal life, the thymus becomes populated by functionally immature (stem) lymphocytes arising from the yolk sac, liver and spleen. In late fetal life, bone marrow becomes the source of these lymphocytes. The transformation of immature lymphocytes to T cells is promoted by a humoral factor called thymosin. Thymosin is believed to be produced by epithelioreticular cells.

Within the thymus, there is large-scale proliferation of cells and large-scale cell death. Most newly formed lymphocytes die within a few days. Nevertheless, large numbers of T lymphocytes are released and carried by the blood to the lymph nodes, spleen, and other lymphatic organs and tissues, where they occupy specific areas. Under proper stimulation, the T cells proliferate and participate in cell-mediated immune response. T cells survive for long periods and recirculate through lymphatic tissue.

Structure of Thymus

Each thymic lobe is composed of thousands of lobules, each of which has an outer cortical and inner medullary component. A capsule encloses each lobe, and septa extend from it as far as the medulla to delineate the lobules. The lobules are not completely separate, because the medullary components of different lobules are continuous with one another. The capsule and septa contain blood vessels, efferent lymph vessels (but no afferent lymph vessels), and nerves.

Lymphocytes are the predominant cell type in the thymus, but plasma cells, granulocytes, mast cells, and fat cells are also present. Macrophages are numerous, especially in the cortex. The framework for the thymic parenchyma is provided by epithelioreticular cells. These cells arise from endoderm, unlike the reticular cells of other lymphatic organs, which arise from mesenchyme. During development, the epithelium of the thymus (which arises as a paired ventral outgrowth from the third pharyngeal pouch) invaginates into and soon obliterates the lumen. Stem lymphocytes invade and occupy the spaces between the epithelial cells. Some of these epithelial cells give rise to the epithelioreticular cells. Despite the large numbers of lymphocytes, the epithelioreticular cells remain joined to one another by desmosomes at the end of long cytoplasmic processes. As a consequence, they assume a stellate shape. They posess ovoid, pale-staining nuclei with a distinct nucleolus.

The epithelioreticular cells correspond to the reticular cells and their associated reticular fibres in other lymphoid tissue in providing a framework. However, the reticular cells of the thymus do not produce any reticular fibres.

Unlike other lymphoid organs, the lymphatic tissue of the thymus is not arranged in nodules. The outer portion, or cortex, of each thymic lobule contains a high concentration of small lymphocytes, also called thymocytes. The inner portion, or medulla, contains a lesser concentration of lymphocytes. In addition to thymocytes, medium and large lymphocytes may also be present. Thus the cortex appears dark-staining, and the medulla appears lighter staining. Because of the density of lymphocytes in the cortex, it is almost impossible to identify reticular cells there; they can sometimes be seen in the medulla. In some tissue sections, the continuities between the medullas of different lobules can also be seen.

Structures called Hassall's corpuscles are a distinctive feature of the thymus. Hassall's (or thymic) corpuscles are isolated masses of concentrically arranged epithelioreticular cells. Their centre may display evidence of keratinization. They are acidophilic and become more prominent during periods of intense destruction of thymocytes and during involution. They are more common in the medulla.

The Blood-Thymic Barrier

Arteries entering the thymus branch and travel along the trabeculae. Arterioles enter the lobules at the junction of the cortex and medulla. Numerous capillary branches pass into the cortex, and less regular branches pass into the medulla. In the medulla, blood vessels carry with them a sheath of loose connective tissue. In the cortex, the perivascular connective tissue sheath becomes ensheathed in epithelioreticular cells. A basement membrane is frequently interposed between the CT and the epithelioreticular cells. These various layers form a blood-thymus barrier in the cortex. From the lumen outward, the layers of the blood-thymus barrier are: the capillary endothelium, the basement membrane of the endothelium, a thin perivascular CT sheath containing many macrophages, the basement membrane of the epithelioreticular cells, the epithelioreticular cell sheath. The endothelium itself has been shown to be highly impermeable to macromolecules. The capillaries in the medulla drain into venules, which also receive the capillaries returning from the cortical zone. (Capillaries are the only kind of blood vessels in the cortical zone.) The medullary veins enter the connective tissue septa and leave the thymus through the capsule. There are no afferent lymph vessels in the thymus, which therefore does not act as a lymph filter, as do lymph nodes. Efferent lymph vessels are few, and are located in the walls of blood vessels and in the CT of the septa and capsule.

Histology of the Heart

The major part of the heart is constituted of cardiac muscle. This kind of muscle is found only in the heart and in the tunica media of the terminal parts of the venae cavae as they enter the heart. The layer of the heart consisting of cardiac muscle is called the myocardium. Its inner surface of the myocardium is lined with endocardium, and the outer surface with epicardium.

- The Myocardium
- Longitudinal section of cardiac muscle
- Branching in cardiac muscle fibres
- Cross section of cardiac muscle

- Longitudional section of Purkinje fibers
- Cross section of Purkinje fibres
- Regular cardiac fibres in cross section
- The Endocardium
- Endocardium of atrium
- Endocardium of ventricle
- The Epicardium
- Low power view of epicardium
- High power view of epicardium.

The Myocardium: Cardiac muscle is similar to skeletal muscle in many ways. Both types appear striated as a result of the arrangement of the actin and myosin filaments in the sarcomeres of the muscle fibres. The sarcolemma (plasma membrane) of both types has invaginations called T-tubules that spread depolarization throughout the cell. The signal is transmitted to the terminal cisternae of the sarcoplasmic reticulum, which are adjacent to the T-tubules. The release of calcium from the cisternae allows contraction to take place. The mechanism of muscle contraction (sliding filament) is the same in the cardiac and skeletal muscle.

There are differences in cardiac and skeletal muscle observable at the light microscope and ultrastructural level. Cardiac muscles fibres are of smaller diameter (about 15 micrometers) than most skeletal muscle fibres (10-100 micrometers). Cardiac muscle fibres are formed by individual muscle cells with one or two centrally placed nuclei, while skeletal muscle fibres are multinucleated protoplasmic units in which the nuclei are peripherally located.

Cardiac muscle fibres branch and anastomose, skeletal muscle fibres do not. Cardiac muscle fibres are arranged in a linear array, each fibre is about 85-100 micrometers long. The junction between two cardiac muscle cells, called an intercalated disk, is another distintguishing feature. The intercalated disk is made up of three types of cell junctions: fascia adherentes, desmosomes and gap junctions.

At the ultrastructural level, the arrangement of T-tubules is more regularly organized in skeletal muscle, and they are found at the A-I junction, in contrast to the Z-line in cardiac muscle. T-tubules are usually associated with two terminal cisternae (triad) in skeletal muscle, versus one (diad) in cardiac muscle.

The cisternae of skeletal muscle are much larger than those of cardiac muscle. Cardiac muscle is more vascularized and has more abundant mitochondria than does skeletal muscle (40% of volume vs. 2%), it also contains glycogen granules between the myofibrils.

Physiologically, cardiac muscle is intrinsically rhythmic (contracts without outside stimulation) although it is regulated through nervous and hormonal mechanisms. The rate of cardiac muscle contraction is set by the sinoatrial node, whose intinsic rhythm is the most rapid.

Histology of the Blood Vessels

The blood vessels are made of three layers, called from the luminal side outward, the tunica intima, the tunica media and the tunica adventitia. These three layers are analogous to the endo-, myo- and epicardium, respectively. The thickness of these three layers varies greatly depending upon the size and type of vessel (large, medium & small arteries and veins; capillaries).

- The tunica intima consists of an endothelium (present in all vessels) and any subendothelial connective tissue that may be present (highly variable depending on vessel). The endothelium of vessels entering or leaving the heart is continuous with that of the heart.
- The tunica media is the layer of concentrically-arranged smooth muscle, the autonomic control of which can alter the diameter of the vessel and affect the blood pressure. Smooth muscle cells (in contrast to cardiac and skeletal) have secretory capabilities, and (depending on the vessel), the tunica media contains varying amounts of collagen fibres, elastic fibres, elastic lamellae, and proteoglycans secreted by the smooth muscle cells. The tunica media of arteries is larger than that of veins of similar size.
- The tunica adventitia is made chiefly of longitudinally arranged collagen fibres. It tends to be much larger in veins than arteries.

Arteries: Arteries carry blood away from the heart. They are classified into three types according to their size: large or elastic arteries; medium (or muscular or distributive) arteries; and small arteries or arterioles, which are less than 0.5 mm in diameter. These types are all continuous with one another. A characteristic feature of arteries, regardless of size, is a well-defined lumen, rounded or oval, maintained by the muscularity of the vessel wall.

Large Arteries: The aorta and its branches (brachiocephalic, subclavian, pulmonary, beginning of common carotid and iliac) are distinguished by their great elasticity. This helps them smooth out the large fluctuations in blood pressure created by the heartbeat. During systole, their elastic laminae are stretched and reduce blood pressure. During diastole, the elastic rebound helps maintain arterial pressure.

Tunica intima: Large arteries often have a large subendothelial layer, which grows with age or disease conditions (arteriosclerosis). Both connective tissue and smooth muscle are present in the intima. The border of the intima is delineated by the internal elastic membrane. The internal elastic membrane may not be conspicuous because of the abundance of elastic material in the tunica media.

Tunica media: This is the thickest of the three layers. The smooth muscle cells are arranged in a spiral around the long axis of the vessel. They secrete elastin in the form of sheets, or lamellae, which are fenestrated to facilitate diffusion. The number of lamellae increase with age (few at birth, 40-70 in adult) and with hypertension. These lamellae, and the large size of the media, are the most striking histological feature of elastic arteries. In addition to elastin, the smooth muscle cells of the media secrete reticular and fine collagen fibers and proteoglycans (all not identifiable). No fibroblasts are present.

Tunica adventitia: This is a relatively thin connective tissue layer. Fibroblasts arethe predominant cell type, and many macrophages are also present. Collagen fibres predominate and elastic fibres (not lamellae) are also present. The collagen in the adventitia prevents elastic arteries from stretching beyond their physiological limits during systole. Blood vessels supplying the adventitia and outer media are also present, these are called vasa vasorum ("vessels of the vessels"). (The inner part of the media is supplied from the lumen via pinocytic transport).

Medium Arteries: The majority of named arteries are medium (muscular or distributive) arteries. There is no sharp dividing line between elastic (large) and muscular (medium) arteries; in areas of transition, arteries may appear as intermediates between the two types. Medium arteries have less elastic tissue than large arteries, the predominant constituent of the tunica media is smooth muscle.

Tunica intima: The tunica intima is thinner than in large arteries, there are fewer smooth muscle cells and less elastic tissue. The outermost part of the intima is defined by a very prominent internal elastic membrane (not obscured by elastic lamellae as in large arteries). The basement membrane of the endothelium may rest directly on the internal elastic membrane, or be separated by a subendothelial layer of CT. The tunica intima increases in thickness with age, and may also become expanded by lipid deposits.

Tunica media: Smooth muscle cells predominate in the tunica media, and little elastic material is present. As in large arteries, no fibroblasts are present. Elastic fibres (few), collagen, and ground substance are produced by the smooth muscle cells. These are arranged in a spiral fashion and their contraction helps maintain blood pressure. In tissue preparation, the internal elastic membrane of the intima appears wavy due to the contraction of the smooth muscle of the media.

Tunica adventitia: The main constituent of the adventitia is collagen fibres, secreted by fibroblasts. Elastic fibres are also present, a concentration of such fibres at the inner boundary of the adventitia is called the external elastic membrane. The external elastic membrane is not as prominent as the internal, and as arteries get smaller disappears much earlier. The tunica adventita is relatively larger than in elastic arteries, it can be up to the same size as the media. It will often blend in with the CT of surrounding structures. Adipose cells may be present.

Small Arteries: Different textbooks will give slightly different numbers for what constitutes the diameter of small arteries and arterioles. Again there is a gradation to larger vessels. The general construction of small arteries is very similar to that of muscular arteries. The media is still muscular and has up to 8-10 layers of smooth muscle cells. This number is reduced as the arteries get smaller, the smallest arterioles have 1-2 layers of smooth muscle cells. The adventitia becomes thinner and the external elastic membrane disappears. The intima becomes smaller and the internal elastic membrane also eventually disappears. However, it persist much longer than the external, and it is not uncommon to see very small arteries which still have an internal elastic membrane. Small arteries also maintain their shape, and tend to be round or oval.

Veins: Veins are the vessels that return blood to the heart. Like arteries, they are classified as large, medium and small, and the sizes blend into one another with no sharp demarcations. Although the same layers (intima, media and adventitia) are present, they are often not as well defined as in arteries. A big difference between arteries and veins is the thickness of their walls and the relative amount of muscle tissue (media). In comparably sized vessels, arteries have thicker walls and a much larger media. In veins, the adventitia is larger than the media. Because of these features, veins do not retain their shape. They often appear floppy in sections, and the lumen may not be patent. Veins are frequently of an irregular shape. Veins also have less elastic tissue than do arteries. Even in larger veins, the internal elastic membrane may be poorly developed or absent. (It may also be present, so don't assume something is an artery just because you see elastic tissue.) Valves, which function to prevent the backflow of blood especially in the lower part of the body, are also seen quite frequently in veins. Veins often travel in close proximity to their arterial counterparts, which is convenient for histological comparison.

Large veins: The large veins are the venae cavae and portal vein and their tributaries. The tunica intima consists of the endothelial lining with its basement membrane, a small amount of subendothelial connective tissue and some smooth muscle cells. It blends in with the tunica media which is relatively thin, and in addition to smooth muscle cells may contain collagen fibres and some fibroblasts (in contrast with the media of arteries). The most distinguishing feature of large veins is the large tunica adventitia. The adventitia is the thickest layer in large veins and is made of collagen fibres, some elastic fibres and fibroblasts. Prominent bundles of longitudinally-arranged smooth muscle are a distinguishing feature.

Medium Veins: The tunica consists of the endothelium and a thin subendothelial layer with smooth muscle cells among the connective tissue elements. A thin internal elastic membrane may or may not be present. (If present, it is not nearly as prominent as in arteries).

The tunica media is much thinner relative to that of an artery, and consists mostly of circularly arranged smooth muscle but also contains collagen fibres. The tunicas intima and media therefore tend to be less distinct from one another than is the case in arteries.

The tunica adventitia is usually thicker than the media and is made up mostly of collagen fibres. It may contain longitudinally oriented smooth muscle bundles. (Remember gradations between the vessels of different sizes are continuous.)

Small Veins or Venules: Different types of venules are described but are not distinguishable with the light microscope. Post capillary venules (which receive blood from capillaries) have only an endothelial lining (intima) and lack a smooth muscle media. They are surrounded by pericytes, which are undifferentiated mesenchymal cells. The basement membrane of the endothelial cells and pericytes may fuse. It is at the level of post-capillary venules that white blood cells leave the blood to enter the tissue. The endothelium of post-capillary venules is the main site of action for vasoactive agents such as histamine and serotonin which cause extravasation of fluid and WBCs during inflammation or allergic reactions.

Collecting venules have a thin adventitia in addition to the pericytes surrounding the intima. The adventitia consists of of longitudinally arranged collagen fibres with a few elastin fibres.

Capillaries: Capillaries are the smallest diameter vessels and the site of exchange of metabolites between blood and tissues. Capillaries are just wide enough to allow the passage of red blood cells, only one cell at a time. (It is during this squeezing through the capillaries that mis-shapen RBCs, as in sickle cell anemia and thalassemia, rupture.)

Capillaries consist of a single layer of endothelial cells and their basement membrane. The endothelial cells are joined together by tight junctions. At intervals, these tight junctions are interrupted, leaving small spaces allowing the passage of fluid between blood and ECF. These interruptions do not occur in the brain, and the lack thereof is responsible for the blood-brain barrier present in most of the brain. Endothelial cells also have pinocytotic vesicles which are involved in transporting macromolecules.

Capillaries are classified according to the structure of their endothelial cells.

1. Continuous capillaries (most capillaries) have a continuous endothelial cells with no fenestrations (openings) in their walls. They are found in nervous tissue, muscle tissue, lung, connective tissue and exocrine glands.
2. a. Fenestrated capillaries have endothelial cells in which are found small openings, called fenestrae, of about 80- 100 nm in diameter. The fenestrations are covered by a small non-membranous diaphragm (which may be the remnant of the glycocalyx enclosed by pinocytotic vesicles from which the fenestrae may be formed). The basement membrane of endothelial cells is continuous over the fenestrae. Fenestrae allow greater permeability and the rapid passage of macromolecules smaller than plasma proteins. Fenestrated capillaries are found in the intestine and endocrine glands. b. A special type of fenestrated capillary with no diaphragm is

found in the renal glomerulus. This capillary has a thick basement membrane.

3. Sinusoids, also called discontinuous capillaries, have a large lumen and follow a tortuous path. They have many fenestrations with no diaphragm, and a discontinuous or absent basal lamina. The lumen is lined with phagocytic cells. They are found in the liver, hematopoietic organs (bone marrow, spleen) and some endocrine organs.

At intervals, the endothelial cells of capillaries may be surrounded by pericytes, undifferentiated mesenchyme cells (as in post-capillary venules). Pericytes can differentiate into fibroblasts and smooth muscle cell, and function in wound repair and the formation of new vessels. Pericytes are associated much more frequently with continuous than with fenestrated capillaries. It is not possible to distinguish different types of capillaries (except sinusoids) with the light microscope. They are recognized by their small (1 RBC) diameter and thin wall.

The Respiratory System

The respiratory system consists of a conducting portion and a respiratory portion. The conducting portion provides a passageway for air and functions to condition the incoming air, by warming, moistening and cleaning it. It consists of the nasopharynx, larynx, trachea, bronchi, bronchioles and terminal bronchioles. (We omit the nasopharynx and larynx in this block.) The respiratory portion serves to rid the body of carbon dioxide and pick up oxygen. It consists of respiratory bronchioles, alveolar ducts and alveolar sacs. All of these structures bear alveoli, the tiny air sacs in which the gas exchange takes place.

Conducting Portion: The conducting portion not only brings the air to the respiratory portion, but also cleanses, moistens and warms the incoming air. Hair and secretions of sebaceous glands in the nasal cavity trap particulate matter, and a countercurrent flow between the incoming air and the blood warms the air. In the trachea and lower portions, secretions of mucous and serous glands moisten the air, which protects the alveoli from desiccation, and trap extraneous matter. These secretions are then directed toward the pharynx by the beating of cilia on epithelial cells. The vasculature in these structures also warms the incoming air.

Trachea: Low power view of human trachea: The trachea is a flexible tube that extends from the larynx into the thorax. Its main function is to act as a conduit for air, but it also helps condition the inspired air. The trachea consists of four layers: a mucosa (epithelium and lamina propria), a submucosa, a fibrocartilage layer and an adventitia. The mucosa consists of a pseudostratified, ciliated, columnar epithelium, (described below) with a very elastic lamina propria. Without special stains, the elastic fibres in the lamina propria (or anywhere else) are not identifiable. The basement membrane of the epithelium appears unusually thick because a large number of collagen fibres are found underneath the actual basement membrane.

The submucosa consists of loose connective tissue with numerous mixed (serous and mucous) glands. Ducts from these glands open toward the lumen. Particles get entrapped in the mucus which floats on the serous secretions. The beating of cilia of epithelial cells directs the secretions toward the oral cavity, where they can be swallowed or spat out. The boundary between the mucosa and the submucosa is not obvious. Special stains would reveal an abundance of elastic fibres at this boundary.

The fibrocartilage layer consists of a series (16-20) of C-shaped cartilage rings which prevent the trachea from collapsing. As is typical of cartilage, the rings are surrounded by a band of dense connective tissue called a perichondrium. The perichondrium merges with the submucosa and the adventitia. At the back, the open ends of the C are joined by a band of smooth muscle called the trachealis muscle. Contraction of this muscle reduces the tracheal diameter and increases intrathoracic pressure during coughing. The area between the rings is occupied by fibroelastic connective tissue. The adventitia is a layer of connective tissue that binds the trachea to adjacent structures in the neck and mediastinum. It contains the largest blood vessels, nerves and lymphatics.

Low Power View of Rabbit Trachea

There is a tear (artifact) that runs along most of the cartilage. The rabbit trachea is in your collection so you can look at the epithelium. (The epithelium on the human trachea slides is in very bad shape.) The most abundant cell type is the ciliated columnar cell. These cells have about 300 cilia at their apical surface, and the cilia sweep in a coordinated fashion from the deepest passageways to the pharynx, to protect the lungs from particulate matter. The cilia appear as a fuzzy line along the top surface of the epithelium. The nuclei of ciliated cells are relatively pale and lie in about the middle of the cell.

The next most abundant cell type is the goblet cell, which secretes mucus. Goblet cells are interspersed among the ciliated cells and also extend the full length of the epithelium. The mucinogen granules are found in the cytoplasm at the apical end. The thick mucus extends the apical end, rendering the cell wineglass-shaped (hence its name). The nucleus is flattened at the base of the mucus cup (not generally identifiable). With special preparation, the mucus can be seen; in standard sections it is washed out leaving a clear (white) area where the mucus had been. Your sections are cut thicker than the width of a single cell, therefore (i) you often just see the top (white) part of the goblet cell, and (ii) the cell might look like it is ciliated, as is the case in the goblet cells seen here. The cilia belong to ciliated cells lying below the goblet cell in the plane of focus. (With the fine focus knob of your microscope, you can focus on cells at different levels and see that the section is quite thick. For this reason, it is also hard to make out the boundaries of any individual cell.)

The third most abundant cell type is the basal cell. Basal cells are short, rounded cells with densely staining nuclei which lie in a row close to the basement membrane. These cells are reserve cells that can differentiate into other cell types.

There are two other cell types in the epithelium that you will not be able to distinguish in standard preparations. One is the small granule cell, which resembles the APUD (amine precursor uptake and decarboxylation) cells of the gut. The granules of these cells can only be seen with special techniques such as silver staining. They secrete catecholamines or polypeptide hormones and may function in regulating the caliber of airways or blood vessels, or be involved in regulating mucous and serous secretions. They resemble basal cells and are found in the same region. However, they are much less abundant, therefore most of the basal nuclei you see will belong to basal cells.

The final cell type is the brush cell. Brush cells are columnar cells with short microvilli at their apical surface (hence name), and their basal surface is in contact with afferent nerve fibres. They are thought to be sensory receptors.

Summary of epithelial cell types: You should be able to identify ciliated columnar cells, goblet cells and basal cells, and know about small granule.

High Power View of Epithelium of Human Bronchus

Because of the poor condition of the epithelium in slide 15 (human trachea). The epithelium is similar to that of the trachea, but the columnar cells are not as tall, as the height of the epithelium decreases continuously as you travel down the respiratory tree. Ciliated columnar cells and goblet cells can be seen, as can some (probable) basal cell nuclei. While you can see good examples of the three identifiable cell types, is not possible to attribute every nucleus seen to a particular type of cell.

Duct of Submucosal Gland

The ducts of the glands pass through the lamina propria and epithelium to empty into the lumen. The lumen is obscured as it approaches the epithelium. (This is a frequent phenomenon of sectioning). However, you can note the abundance of lymphocytes (appear as purple dots) in the connective tissue of the lamina propria.

High Power View of Cartilage

The matrix of cartilage typically stains a purplish colour. Staining is most intense around the cartilage cells, which are called chondrocytes and sit in spaces called lacunae. Chondrocytes are often found in clusters (called isogenous groups). During life, the cells occupy the whole lacuna, but they frequently shrink during preparation, and the lacunae appear as spaces around cells (or as empty spaces if a cell is lost). The cells at the periphery of the cartilage are smaller and more elongated than centrally located cells. Cartilage is surrounded by a perichondrium, a band of dense connective tissue. The perichondrium blends with the connective tissue of surrounding structures (here the adventitia; the extensive spaces in the CT are artifacts of preparation). The details of the histology of cartilage are described in the Musculoskeletal Block. For the purposes of this block, you are only expected to recognize it.

Lower Power View of Trachealis Muscle and Glands

A lower power view of the human trachea at the area of the trachealis muscle is shown. The trachealis muscle consists of bands of smooth muscle whose fibres insert into the perichondrium of the cartilage and connect the two ends of the C. Contraction of this muscle reduces the tracheal diameter and raises intrathoracic pressure during coughing. Mixed glands are found interspersed within and around this muscle.

Bronchi: The end of the trachea divides into two primary bronchi that enter the lungs. Histologically, the trachea and primary bronchi are very similar and you will not be expected to distinguish between them. Within the lung, the primary bronchi almost immediately divide into secondary (or lobar) bronchi that supply each lobe. Secondary bronchi divide into tertiary (or segmental) bronchi that supply the segments of each lobe. These bronchi continue to divide to give rise to many generations of bronchi, each smaller than the preceding one. You are not expected to distinguish different generations of bronchi, you are only expected to tell a bronchus from the trachea and from bronchioles.

In bronchi, the C-shaped cartilage of the trachea is replaced by separate plates of cartilage. These plates become smaller and farther apart the more distal the bronchus. (When no more cartilage is present, you have a bronchiole, not a bronchus). At the same time, the lamina propria becomes surrounded by a band of smooth muscle. The muscle fibres that appear in the bronchi are arranged spirally and criss-cross one another, so they do not appear as a continuous band in a cross-section. This band of smooth muscle becomes a more conspicuous feature as the cartilage diminishes. The smooth muscle can be considered as a separate layer, the muscularis, lying between the mucosa on the one side and the submucosa, fibrocartilage layer (with plates) and adventitia on the other. Going distally down the bronchial tree, the lamina propria becomes reduced and the smooth muscle of the muscularis comes to lie closer to the epithelium. Serous and mucous glands are present in the submucosa of bronchi, their numbers also decrease with each division into smaller orders of bronchi. (Like the cartilage, all glands are gone when bronchi branch into bronchioles.) The epithelium is similar to that of the trachea, but its height becomes reduced and goblet cells become less frequent.

About blood vessels in the lungs: Blood vessels are frequently seen next to components of the respiratory tree. That is because branches of the pulmonary artery travel along the respiratory tree to become distributed in a capillary network over the alveoli, where gas exchange occurs. In addition to branches of the pulmonary arteries, whose purpose is to bring deoxygenated blood to the lungs for oxygenation, bronchial arteries arising from the aorta are present. Bronchial arteries are nutrient vessels that supply the tissue of the lung with oxygen. They also accompany the respiratory tree, and at the level of respiratory bronchioles, anastomose with pulmonary arteries. Veins arising from the capillary network around the alveoli can also be seen, they are extremely thin-walled. These veins travel independently of arteries, that is they will not follow the respiratory tree but are heading toward the intersegmental connective tissue where they will join to form the pulmonary veins. Most of the blood reaching the lungs via the bronchial arteries is drained by the pulmonary veins. Bronchial veins drain only the the connective tissue in the hilar region. Because the right side of the heart operates at a much lower pressure than the left, pulmonary arteries tend to have much less muscular walls than do typical arteries. However, they contain a great deal of elastic tissue and are considered elastic arteries. Bronchial arteries are smaller, but their structure is more typical of regular arteries, that is they have a relatively thicker tunica media. The branches of the pulmonary veins will have extremely thin walls. You will not be asked to distinguish different types of vessels in the lung, but you should be aware of the functions of the pulmonary and bronchial circulation.

Wall of a more Distal Bronchus

The epithelium is has become shorter (it is still pseudostratified, columnar ciliated) and goblet cells are less abundant. The muscularis has come to lie very close to the epithelium and the lamina propria is reduced to a thin band underneath the epithelium. The submucosa is also very scanty. A large cartilage plate lies almost directly under the muscularis. In this bronchus, there would be fewer cartilage plates and fewer glands in the submucosa.

Bronchioles: As the bronchi branch and become smaller, the cartilage plates become smaller and farther apart and the mixed glands become fewer. When the cartilage and glands have disappeared, you have entered the bronchioles. At this point, the muscularis is very close to the epithelium. The epithelium is still pseudostratified, ciliated columnar, although it becomes progressively less tall. Goblet cells are still quite common in large bronchioles, their numbers also progressively decrease.

Terminal Bronchioles

As the (regular) bronchioles branch into smaller and smaller bronchioles, they eventually give rise to terminal bronchioles. As their name implies, terminal bronchioles represent the last part of the conducting portion of the respiratory tree. In terminal bronchioles, the ciliated pseudostratified epithelium abruptly gives way to simple cuboidal epithelium consisting of Clara cells. Clara cells are secretory: the lipoprotein they secrete prevents luminal adhesion during expiration and inactivates harmful substances.

Respiratory bronchioles: The purpose of the conducting part of the respiratory tree is to get the air to the millions of tiny air sacs, or alveoli (sing. alveolus) of the lungs, where the gas exchange occurs. The respiratory portion of the tree begins in the respiratory bronchioles when these alveoli first arise. The respiratory portion begins when an air sac appears as an outpocketing of the bronchiole. At this point, the terminal bronchiole has become a respiratory bronchiole.

Alveolar Ducts, Alveolar Sacs and Alveoli

As the respiratory bronchioles continue and branch, the alveoli forming outpocketings from their walls become progressively more frequent and closer together. Eventually alveoli are so frequent that they are side by side separated only by their walls (alveolar septa). This marks the beginning of the alveolar ducts. At first, the ends of the septa facing into the lumen of the alveolar duct have prominent knobs of smooth muscle. These become progressively less prominent and eventually disappear.

Septa of the alveoli: The septa between alveoli are specialized for the diffusion of gases. The surfaces facing the air are lined by an epithelium made of two types of cells, type I and type II alveolar cells. Most of the surface (about 95%) is lined by type I alveolar cells (or type I pneumocytes). These cells are extremely squamous. The other type of epithelial cell is the type II alveolar cell (or type II pneumocyte, or great septal cell). Type II cells are found interspersed among the type I cells, singly or in clusters. They are often found in the angles of the alveolar walls. Type II cells secrete surfactant, a phospholipid that spreads over the alveolar surfaces to reduce surface tension. This prevents the alveoli from collapsing upon expiration. (Premature babies with insufficient surfactant suffer from hyaline membrane disease.) Surfactant also has a bacteriacidal effect. Type II cells are fairly large cuboidal cells that have a large nucleus and vacuolated cytoplasm. They are about as abundant as type I cells even though they cover only 5% of the surface area of the alveoli.

Within the septal wall are found capillaries (the alveoli contain the richest capillary network of the body), elastic and collagen fibres, fibroblasts and macrophages called dust cells. The septal walls have thick portions and thin portions. The thick portions could contain any of the components listed above. The exchange of gases occurs in the thin portions of the septa, which contain only the following three components:

(1) the cytoplasm of the type I cell, (2) the fused basement membranes of the type I cell and the endothelial cell, and (3) the cytoplasm of the endothelial cell of a capillary.

When you look at the alveolar septa, you will be able to identify (some of the) type II cells, capillaries if well-sectioned, and (if they are filled with carbon particles or some other material) dust cells. You will not be able to tell the nuclei of type I cells from those of fibroblasts. There is no point in trying to identify every cell in the septal walls.

The Urinary System

The urinary system consists of the paired kidneys, the paired ureters, which lead from the kidneys to the bladder, the urinary bladder, which temporarily stores the urine, and the urethra, which leads from the bladder to the exterior of the body.

The Kidneys

General function: The kidneys are highly vascular, receiving about 25% of the cardiac output. They produce urine, in which various metabolic waste products are eliminated. Urine arises as an ultrafiltrate of blood which is modified by selective resorption and specific secretion by cells in the kidney. It contains water, electrolytes, waste products such as urea, uric acid and creatinine, and the breakdown products of various substances. The kidneys also keep the body's extracellular water content and ionic composition almost constant, even though the dietary intake of water and salt may vary.

Endocrine Function

In addition to their activities in waste removal and homeostasis, the kidneys function as an endocrine organ. Several compounds, listed below, are produced by the kidneys and released into the blood.

- Erythropoetin (EPO). Erythropoetin, a glycoprotein growth factor regulating red blood cell formation, is synthesized and secreted by the kidney in response to decreased tissue oxygen tension. The precise cellular source of EPO is not resolved, but it is believed to be the capillary endothelial cells or the fibroblasts in the peritubular tissue of the inner cortex and outer medulla. Tubular epithelial cells of the inner cortex may also be involved in EPO synthesis.
- Renin. Renin is a hormone involved in the control of blood pressure and volume. It is produced by the juxtaglomerular (JG) cells of the afferent arteriole (described below). The JG cells produce, store and release renin into the blood, where it catalyses the hydrolysis of circulating angiotensinogen to produce the decapeptide angiotensin I. Angiotensin I is converted to the octapeptide angiotensin II by an enzyme in the lung. Angiotensin II is the active agent in blood pressure

regulation and function. It is a potent vasoconstrictor with a regulatory role in the control of renal and systemic vascular resistance. It also stimulates the release of aldosterone from the adrenal gland. Aldosterone increases the resorption of sodium, and the concomitant resorption of water, from the distal tubules, thereby raising sblood pressure and volume.

- 1,25-dihydroxy vitamin D. Vitamin D, despite its name, is a steroid prohormone. It is converted to its highly active dihydroxy form by successive metabolic steps that take place first in the liver and then in the kidney. 1,25-dihydroxy vitamin D plays an essential role in calcium metabolism. It stimulates the intestinal absorption of calcium and is needed for the normal calcification of bone.

The Structure of the Kidney

The kidneys are large, reddish organs located on either side of the spinal column on the posterior abdominal wall. They are covered with a thin but strong capsule of dense connective tissue. The medial border of each kidney is concave with a deep vertical fissure, the hilum, through which renal vessels and nerves pass. The expanded, funnel-shaped origin of the ureter, called the renal pelvis, also leaves through the hilum.

The renal pelvis is made of two or three divisions called the major calyses. Each major calyx branches into several cup-shaped minor calyces. The apical portion-called the papilla- of each renal pyramid projects into a minor calyx. (Renal pyramids are medullary structures which contain nephric tubules. They are described in more detail below.) The tip of the papilla is perforated with the openings of 10-25 papillary ducts, the final collecting ducts of the uriniferous tubules, and is therefore called the area cribrosa.

The renal pelvis is surrounded by a space called the renal sinus. The capsule, which passes inward at the hilum, forms the connective tissue covering of the sinus and becomes continuous with the connective tissue forming the walls of the renal pelvis and minor calyces. The sinus itself is largely filled with loose connective tissue and adipose tissue.

When a kidney is hemisected and viewed with the naked eye, two distinct regions can be seen. The outer part, called the cortex, is a dark reddish brown colour. The inner part, called the medulla, is lighter-colored. The difference in colour reflects the distribution of blood. About 90-95% of the blood passing through the kidney is in the cortex, the remaining 5-10% is in the medulla. The following brief description of the nephron, the functional unit of the kidney, will make it easier to follow the discussion of the appearance of the cortex and medulla.

The Nephron and its Collecting Ducts

Each kidney has about two million nephrons. The components of the nephron are:

1) the renal corpuscle, which consists of a special arterial capillary bed called the glomerulus, and Bowman's capsule, which surrounds the glomerulus and holds the fluid exuded from it. The arteriole leading to the glomerulus is called the afferent arteriole, and the one draining it is called the efferent arteriole.
2) the proximal thick segment, consisting of the proximal convoluted tubule (PCT, also known as pars convoluta) and proximal straight tubule (pars recta). The proximal convoluted tubule arises from Bowman's capsule and remains in the proximity of the renal corpuscle. It continues into the proximal straight tubule, which forms the thick descending limb of the loop of Henle-the loop's first segment.
3) the thin segment, which makes up the thin limb of the loop of Henle-the second segment of the loop. The loop of Henle is a U-shaped structure. Some distance down the descending limb of the U, the proximal thick segment gives way to the thin segment. The bottom of the U, which is in the form of a hairpin turn, may -depending on the nephron-consist of thin segment (in long loops of Henle) or be the beginning of the thick ascending limb (in short loops). In long loops of Henle, the thin segment continues for some distance up the ascending limb of the loop. The parallel arms (descending and ascending) of long loops are responsible for creating the countercurrent exchange mechanism essential for the formation of urine.
4) the distal thick segment, consisting of the distal straight tubule (pars recta) and the distal convoluted tubule (DCT or pars convoluta). The distal straight tubule makes up the third and final segment of the loop of Henle-the thick ascending limb. It continues into the distal convoluted tubule which returns to the vicinity of its own renal corpuscle.

The components of the nephron constitute the secretory portion of the uriniferous tubules. The distal convoluted tubule will connect to the excretory portion of the uriniferous tubules, the collecting tubules or ducts. The collecting ducts run alongside the straight segments of the nephron. The secretory (= nephron) and excretory portions originate from separate embryonic primordia which later become connected.

The Cortex

The cortex contains renal corpuscles as well as convoluted and straight parts of nephric tubules. It also has collecting tubules and an extensive vascular supply.

The cortex is not uniform in appearance. The renal corpuscles, barely discernable with the naked eye, are surrounded by their convoluted tubules, both proximal and distal. The areas of the cortex containing renal corpuscles and convoluted tubules are called cortical labyrinths.

The straight portions of nephric tubules (proximal and distal) as well as collecting tubules, which are also straight, are found in areas called medullary rays. These are a series of vertical striations that appear to emanate from the medulla (hence the name). There are 400-500 medullary rays projecting into the cortex from the medulla.

The Medulla

The medulla is made up of 8 to 18 conical subdivisions called renal pyramids. The pyramids contain the straight segments of the nephrons and the collecting ducts, continuous with those in a group of medullary rays of the cortex. Each pyramid has its base toward the cortex and its apex, or papilla, projecting into a minor calyx. Each pyramid can be divided into an outer and inner zone.

The outer zone (adjacent to the cortex) can be divided into an outer and inner stripe. The zones and stripes reflect the location of distinct segments of nephric tubules at specific levels within the pyramid, and are readily recognized in fresh specimens.

The pyramids are accompanied by a capillary network called the vasa recta that runs in parallel with the various tubules and constitutes the vascular part of the countercurrent exchange mechanism.

The lateral boundaries of each pyramid are defined by inward extensions of the darker cortical tissue forming the renal columns (of Bertin). Although they contain the same components as the cortex, the renal columns are considered to be part of the medulla. A renal pyramid together with the cortical tissue overlying its base and covering its sides constitutes a renal lobe.

Different Types of Nephrons

Several types of nephrons have been identified, depending upon the location of their renal corpuscles within the cortex.

- Subcapsular or cortical nephrons have their renal corpuscles in the outer part of the cortex. They have short loops of Henle that do not go beyond the outer region of the pyramid. The thin segment in these nephrons is confined to a small part of the descending limb, and the distal thick segment begins at the hairpin turn.
- Juxtamedullary nephrons have their renal corpuscles close to the base of a pyramid. They have long loops of Henle with long thin segments. The thick descending limb does not extend beyond the outer stripe of the medulla. The thin limb begins here and extends deep into the inner region of the medullary pyramid. The thick ascending limb begins at a deeper level than that at which the thick descending limb ends. (In other words, the thin ascending limb is shorter than the thin descending limb.) The part of the outer medulla containing the thick segments of both the descending and ascending limbs is called the outer stripe. The inner stripe of the outer medulla contains thin segments of the descending limb, and thick segments of the ascending limb. The inner medulla contains only thin segments, descending and ascending. These differences in the proportion of thick and thin segments account for the colour differences of the zones and stripes that can be seen in sagittal sections of fresh specimens.
- Intermediate nephrons have their renal corpuscles in the midregion of the cortex. Their loops of Henle are intermediate in length. The descending thick segment ends at the outer stripe (just as in the case of the juxtamedullary nephrons). The thin segment, however, makes its hairpin turn within the inner stripe, instead of penetrating into the inner medulla. The thick ascending limb arises within the inner stripe (as in juxtamedullary nephrons), not far beyond the point at which the thin segment comes out of its hairpin turn.

There is a much greater number of shorter loops than of longer loops. Only about one in seven nephrons are of the juxtamedullary kind with long loops of Henle.

Interstitial Tissue

The connective tissue of the kidney parenchyma is called interstitial tissue. It increases in amount from the cortex, where it constitutes about 7% of the volume, to the inner region of the medulla, where it constitutes about 20% of the volume. Two kinds of interstitial cells are found in the cortex. The first kind is found between the basement membrane of the tubules and the adjacent peritubular capillaries. It resembles fibroblasts and secretes the collagen and glycosaminoglycans of the extracellular matrix of the interstitium. The second type of cell is a macrophage.

The principal interstitial cells in the medulla resemble myofibroblasts. They may have a role in compressing the tubular structures. They contain prominent bundles of actin filaments, abundant rough ER, well-developed Golgi, lysosomes and lipid droplets. Prostaglandin and prostacyclin have been reported to be synthesized in the interstitium.

Kidney Lobes and Lobules

Lobulation in the post-natal kidney is not evident externally. A lobe is defined as a single medullary pyramid and the associated cortical tissue at its base, and at its sides (one-half of each adjacent renal column of Bertin). A lobule is defined as a papillary duct (the final collecting duct) and all the cortical collecting ducts and associated nephrons that empty into it. Thus, a central medullary ray and the surrounding cortical material constitutes a lobule. The boundaries between adjacent lobules is not clearly demarcated with connective tissue.

The Blood Supply of the Kidney

Each kidney is supplied by a branch of the aorta called the renal artery. In the renal sinus, it branches into a number of large interlobar arteries. The interlobar arteries pass between the pyramids to the boundary zone of the medulla and cortex. Here they bend sharply and form short arches, the arcuate or arciform arteries, which run along the base of the pyramid between the medulla and cortex. The arcuate arteries divide into a large number of finer branches, the interlobular arteries, which ascend perpendicularly through the cortical labyrinth to the surface, about midway between adjacent medullary rays.

The interlobular arteries give off numerous short, lateral branches, each of which enters a renal corpuscle as an afferent arteriole. A single afferent arteriole may spring directly from an interlobular artery or a common stem from an interlobular artery may branch to form several afferent arterioles. Some interlobular arteries enter the kidney capsule to provide its arterial supply.

The afferent arteriole branches into a tuft of capillaries with 10-20 loops, called the glomerulus. The glomerulus is an entirely arterial capillary bed. It is drained by the efferent arteriole, which leaves the glomerulus near the point at which the afferent arteriole enters. The point at which the two arterioles penetrate Bowman's capsule is called the vascular pole of the renal corpuscle. Opposite it is the urinary pole, where the proximal convoluted tubule arises. Bowman's capsule itself is a double-layered epithelial cup.

The efferent arterioles give rise to another network of capillaries called the peritubular capillaries. The arrangement of these capillaries differs according to whether they arise from glomeruli near the outer surface of the cortex or from the deeper, juxtamedullary glomeruli.

- Efferent arterioles from subcapsular glomeruli lead to a peritubular capillary network that surrounds the local uriniferous tubules.
- Efferent arterioles from juxtamedullary glomeruli descend into the medulla alongside the loops of Henle. They break up into smaller vessels, arteriolae rectae, that continue toward the apex of the medullary pyramids. At various levels, the arteriolae rectae make hairpin turns to return again as straight veins, venulae rectae, toward the base of the pyramid. The descending arterioles and ascending venules are in close proximity to one another and together are called the vasa recta (Latin for straight vessels). The endothelium of the arterial limb is continuous but that of the venous limb is very thin and fenestrated. The proximity of the two limbs and the large surface area they present to one another facilitate the rapid movement of diffusible substances. The vasa recta are an important part of the countercurrent multiplier mechanism involved in concentrating urine.

The arteriolae rectae also give rise to long-meshed capillary nets lined by fenestrated endothelium. They supply the tubular structures at the various levels of the medullary pyramids. (Note that the medulla does not have a direct arterial blood supply: all the blood it receives has first been through the glomeruli in the cortex.)

The venous flow of the kidney follows a reverse course to arterial flow, with the veins running in parallel to the corresponding arteries.

In the cortex, peritubular capillaries drain into interlobular veins, which drain into arcuate veins, interlobar veins and the renal vein. In the medulla, the peritubular vessels (vasa recta) drain into the arcuate veins, and so on. Capillaries from the capsule and peritubular capillaries near the surface drain into stellate veins that enter interlobular veins.

Lymphatics

The kidney has two major networks of lymphatic vessels, which are generally not seen in routine histological preparations. One is located in the capsule and outer region of the cortex and drains into larger lymph vessels of neighbouring organs. The other arises between the uriniferous tubules, especially in the cortex, and enters lymphatics that accompany the larger blood vessels, and leaves through the hilum. There are anastomoses between the two networks.

Functional Histology of the Kidney Tubules

The Renal Corpuscle

The renal corpuscle consists of the glomerulus and the Bowman's capsule that surrounds it. As described above, the glomerulus is a tuft of arterial capillaries composed of 10-20 loops, supplied by an afferent arteriole and drained by an efferent arteriole. This arrangement produces a relatively high filtration pressure.

The cytoplasm of glomerular endothelial cells is thin and has numerous fenestrations. The fenestrations are larger, more numerous and more irregular in outline than those of other fenestrated capillaries. They also lack a diaphragm. Bowman's capsule is an epithelial cup with a visceral (or glomerular) layer and parietal (or capsular) layer. The visceral layer is intimately associated with the glomerular capillaries.

It is made of specialized cells called podocytes ("foot cells"). The parietal layer is made of simple squamous epithelium. A space, called Bowman's space or the urinary space, lies between the two layers. The ultrafiltrate of blood emanating from the glomerulus is collected in this space.

When viewed with the light microscope, the nuclei of the simple squamous epithelial cells forming the parietal layer of Bowman's capsule can easily be identified. The urinary space between the glomerulus and the parietal layer is also obvious. However, the glomerulus itself appears as a large cellular mass. It is not possible to distinguish podocytes from glomerular endothelial cells or other cells called mesangial cells. The description of podocytes and mesangial cells which follows below is based on observations with the electron microscope.

Podocytes

Podocytes cover the glomerular capillaries. Their cell bodies are usually not found in extensive contact with the basement membrane. They stand off by one or two micrometers, but are attached to it via their primary processes, which ramify tentacle-like over its surface.

Each primary process has numerous secondary processes, which in turn give rise to pedicels or foot processes. The pedicels interdigitate with those of neighbouring podocytes. The spaces between the interdigitating pedicels, called filtration slits, are about 25 nm wide and are spanned by an electron dense membrane called the filtration slit membrane. The pedicels contain numerous actin microfilaments, which may have a role in regulating the size and patency of filtration slits.

The endothelial cells of the glomerulus share a basement membrane with the podocytes (visceral layer of Bowman's capsule). The filtration apparatus thus consists of:

- the endothelium of the glomerular capillaries
- a joint basement membrane
- the podocytes.

The filtration apparatus can be described as a semi-permeable barrier. The two cellular layers are discontinuous, while the basement membrane, which is relatively thick, forms a continuous extracellular layer and is the principal component of the filtration barrier. Because of its thickness, it is prominent in histological sections stained with the periodic acid-Schiff (PAS) procedure and therefore also called the glomerular basement membrane (GBM).

The GBM restricts the movement of particles larger than 70,000 daltons, such as proteins like albumin or hemoglobin. Though albumin is not a usual constituent of urine, its occasional presence indicates that it is close to the effective pore size of the filtration barrier. Though the filtration barrier restricts the movement of protein, several grams pass through it each day. This protein is reabsorbed by endocytosis in the proximal convoluted tubule. Polyanionic glycosaminoglycans in the basement membrane restrict the movement of cationic molecules, even those smaller than 70,000 daltons. The narrow slit pores and the filtration slit membranes also act as physical barriers to bulk flow and free diffusion. The fenestrae of the endothelial cells restrict the movement of blood cells and other formed elements of the blood. In addition to structural barriers, the flow rate and pressure of the blood in the glomerulus also have an effect on the filtration function.

The Mesangium

Another group of cells found in the renal corpuscle are mesangial cells, which together with the extracellular matrix they produce constitute the mesangium. The mesangium is most obvious at the vascular stalk of the glomerulus and in the interstices of adjoining glomerular capillaries. Mesangial cells are irregularly shaped, modified smooth muscle cells. One function of the mesangium is to provide structural support for the podocytes and glomerular capillaries. Mesangial cells are also phagocytic, removing trapped residues, aggregated proteins and immune complexes from the GBM. (Atypical for phagocytes, they do not arise from the mononuclear phagocytic system.) Since mesangial cells are capable of contraction, they are thought to be involved in local regulation of glomerular blood flow. They respond to, and release, a number of growth factors and can produce vasoactive and immunoregulatory inflammatory mediators, such as prostaglandins and interleukin 1. Mesangial cells are capable of dividing, and in some disease conditions, mesangial proliferation is stimulated.

The Proximal Thick Segment

The proximal thick segment includes the proximal convoluted tubule (PCT) and the proximal straight tubule. The proximal convoluted tubule arises at the urinary pole of the renal corpuscle. Its lumen is continuous with the urinary space. The simple squamous epithelum of Bowman's capsule changes abruptly to the simple cuboidal epithelium of the PCT. The PCT is the longest and widest part of the nephron, and constitutes the majority of tubular sections seen in the cortical labyrinth. It follows a tortuous course, then terminates by straightening out and passing into the nearest medullary ray to become the thick descending limb of the loop of Henle. The PCT is about 15 mm long, and the entire proximal thick segment can be up to 25 mm long.. The convoluted and straight parts of the proximal tubule have an essentially similar structure and function. The epithelium consists of a single layer of cuboidal cells. The cells of the simple cuboidal epithelium contain a single spherical nucleus in an eosinic cytoplasm. Not all the cells in a given section will have a visible nucleus, because the cells are considerably greater in width than the thickness of the sections.

On their luminal surface, the cells have a conspicuous brush border, which consists of numerous relatively long microvilli. A terminal bar apparatus, consisting of a narrow tight junction and a zonula adherens, seals off the intercellular space from the lumen. Long rodlike mitochondria in the basal half of the cells, oriented parallel to the cell axis, often give this part of the cell a faint striated appearance in the light microscope. The lateral limits of the cell cannot be discerned with the light microscope because their sides are elaborately fluted and interdigitate deeply with complementary ridges and grooves of neighbouring cells. Near the cell base, the electron microscope reveals an even greater number of slender lateral processes extending under adjacent cells. They contain longitudinally-oriented microfilaments which may play a role in regulating the movement of fluid from the basolateral extracellular space across the basement membrane toward the adjacent peritubular capillaries.

Sections of the PCT in the cortical labyrinth often appear irregular in profile (eg. U-, J- or S-shaped) and the lumen is often star-shaped (as opposed to rounded). Within the medullary rays, the tubular profiles of the straight segments are rounded, but many display a star-shaped lumen. Their larger diameter, brush border and usually more eosinophilic staining help to distinguish them from other components of the ray. The straight segment is not as active as the PCT and contains fewer microvilli and fewer, randomly distributed mitochondria.

The PCT reabsorbs about 80% of the primary filtrate. The pump enzyme, sodium-potassium-activated ATPase, actively transports sodium into the lateral intercellular space. This is followed by the passive diffusion of chloride. The accumulation of NaCl creates an osmotic gradient that draws water from the lumen into the intercellular space. The hydrostatic pressure that builds up in the distended intercellular compartments, presumably aided by the contractile activity of the actin filaments at the base of the cell, drives an essentially isosmotic fluid through the basement membrane of the tubule into the renal connective tissue. Here it is reabsorbed into the vessels of the peritubular capillary network. Angiotensin II promotes the resorption of Na in the PCT, and also of HCO3. The PCT also reabsorbs amino acids, glucose and other sugars, and polypeptides. The microvilli of the brush border are covered with a well-developed glycocalyx that contains ATPases, peptidases and disaccharidases. Proteins and large peptides are reabsorbed by endocytosis.

The Thin Segment

The thin segment of the loop of Henle is recognized by its much lower epithelium, which is either simple squamous or low cuboidal. This segment is only about one-quarter the diameter of the proximal tubule. The transition from proximal thick to thin segment is abrupt. With the light microscope, some sections of the thin segment look flatter than others. With the electron microscope, four types of epithelial cells have been identified in the thin segment. (The detail below is for reference only. On your slides, you will only be required to identify a section as "thin segment.")

- Type I is a thin simple epithelium with almost no interdigitations with neighbouring cells and few organelles. It is found in short-looped nephrons.
- Type II is a bit taller (low cuboidal). It has abundant organelles and many small, blunt microvilli. The extent of lateral interdigitation with neighbouring cells shows species variation. It is found in the initial part of the thin descending limb of long-looped nephrons.
- Type III epithelium is thinner and lacks interdigitations. The cell structure is simpler and with fewer microvilli than type II. It is found in the more distal part (within the inner medulla) of the thin descending limb of long-looped nephrons.
- Type IV epithelium consists of low, flattened cells without microvilli and with few organelles. It is found at the bend of the loop and through the entire ascending limb of long-looped nephrons.

The function of the thin segment is discussed below in the section on the countercurrent multiplier mechanism.

The Distal Thick Segment

The distal thick segment consists of the distal straight tubule and the distal convoluted tubule (DCT). The straight tubule is the thick ascending limb of the loop of Henle. It passes through a pyramid (in the long loops) and continues into a medullary ray. From there, it enters the cortical labyrinth and makes contact with the vascular pole of its parent renal corpuscle. At this point, the epithelial cells of the tubule adjacent to the afferent arteriole are modified to form the macula densa (described below). The tubule then leaves the region of the corpuscle to become the distal convoluted tubule. The DCT is about 5 mm long, or one-third the length of the PCT (which is why you see many more profiles of PCT around the renal corpuscles), and follows a less tortuous course. The DCT ends in an arched collecting tubule or a shorter tubule called a connecting tubule, both of which lead to a straight collecting tubule in a medullary ray.

Distal tubules are lined with cuboidal cells that stain lightly with eosin but are less acidophilic than the cells of the proximal tubules. Because the cells are flatter and smaller than those of the proximal tubule, more cells and more nuclei are seen in a typical profile. The lateral margins of the cells are indistinct in the light microscope because the cells have extensive basolateral plications that interdigitate with those of other cells. Numerous large mitochondria associated with folds in the basal membrane sometimes gives the basal part of the cells a striated appearance (again not as prominent as in proximal tubules).

The nucleus is located in the apical portion of the cell. Microvilli are present on the apical surface but they are much fewer in number and shorter than those of the proximal tubule. Consequently, no brush border is evident with the light microscope and the luminal surface of distal tubule sections appears cleaner and sharper than that of proximal tubules.

The outside diameter of distal tubules is slightly smaller than that of proximal tubules. In the cortical labyrinth, profiles of DCTs are rounder than those of PCTs, which, as described above, are often irregular. In the medullary rays, where all the tubules are straight, all the profiles are round.

The distal tubule is responsible for the conservation of Na. Its cells have Na-K-ATPase on their basolateral surface, and Na resorption is coupled with K secretion. The luminal domain of the cell membrane has a cotransporter, driven by the inwardly directed Na gradient, that couples inward Cl transport to the Na and K transport. The resorption of Na and secretion of K is stimulated by aldosterone from the adrenal gland. The DCT also resorbs bicarbonate ion and concomitantly secretes hydrogen ion, causing an increase in urine acidification.

In order to avoid the toxic effects of ammonia, it converts it to ammonium ion that can enter the urea cycle. The terminal portion of the DCT is sensitive to antidiuretic hormone (ADH, also called vasopressin), which increases the permeabiltiy of the tubule to water, to produce a more concentrated urine. (This effect also occurs in the collecting ducts, discussed below.) ADH is released from the posterior pituitary in response to increased plasma osmolality or decreased blood volume. The thick ascending limb is also involved in the countercurrent multiplier mechanism.

The Juxtaglomerular Apparatus

As the distal tubule contacts its parent corpuscle, it passes through a crevice between the afferent and efferent arterioles. The epithelial cells of the tubule adjacent to the afferent arteriole are modified to form the macula densa. The cells of the macula densa have a distinctive appearance. They are narrower and usually taller than other distal tubule cells, and their nuclei appear crowded, even superimposed (hence the name macula densa = dense spot). The cells of the afferent arteriole at this point are also modified, and are called juxtaglomerular cells (JG cells). The JG cells and the macula densa, as well as some associated mesangial cells, together constitute the juxtaglomerular (JG) apparatus.

The JG cells are modified smooth muscle cells of the tunica media of the afferent arteriole. They have ellipsoid nuclei and cytoplasm full of granules that stain with PAS. The electron microscope reveals that they have the characteristics of protein-secreting cells. They produce the enzyme renin. As described above, renin converts the blood protein angiotensinogen into an inactive decapeptide called angiotensin I (AI), which in turn is converted to the active octapeptide angiotensin II (AII) by endothelial cells in the lung.

Renin secretion increases if the NaCl concentration decreases. The cells of the macula densa monitor NaCl concentrations in the afferent arteriole, and regulate the release of renin from the JG cells in a paracrine manner. A significant blood loss or a drop in blood pressure from other causes also stimulates renin secretion, probably as a result of decreased tension in the wall of the JG cells. AII is a potent vasoconstrictor, and therefore directly increases blood pressure. The main physiological effect of AII appears to be to enhance the secretion of the adrenocortical hormone aldosterone. Aldosterone acts mainly on the distal tubules to increase the resorption of Na and Cl ions. This leads to the expansion of the extravascular fluid volume, and an increase in blood pressure. The increased blood pressure distends the JG cells and inhibits renin secretion, completing the negative feedback loop.

The cells of the macula densa also lie adjacent to some extraglomerular mesangial cells, called lacis cells, at the vascular pole of the renal corpuscle. Lacis cells are coupled to one another and to JG cells by gap junctions. They are believed to be involved in transmitting information from the macula densa to JG cells. The amount of renin released is probably determined through a functional cooperation of the macula densa, JG cells and lacis cells.

The Collecting Ducts

A distal convoluted tubule in the cortical labyrinth connects to a straight collecting tubule lying within the medullary ray either via an arched collecting tubule or a connecting tubule. A straight collecting tubule receives input from seven to ten nephrons. From the medullary ray, the collecting tubules pass inward through the outer zone of the medulla without further fusions. When they reach the inner zone, they join at acute angles with other tubules. There are about seven successive convergences in the inner medulla, which result in the formation of large, straight ducts called the papillary ducts of Bellini. The papillary ducts open on the area cribrosa at the apex of each papilla. (Note: Some people use the term "tubule" for the smaller structures, and "ducts" for the larger ones, other people use the terms interchangeably.

We will accept either term. What is important is that you recognize a structure as a collector.). The collecting ducts are lined with simple epithelium throughout their length. The arched collecting tubules and connecting tubules are lined with flattened cells, squamous to cuboidal in shape. (Do not spend time looking for them, concentrate on finding the collecting ducts in the medullary rays and medulla.) The epithelium of the straight collecting tubule in the cortex is cuboidal, but becomes taller as the duct progresses through the medulla. The papillary ducts have tall columnar epithelium. The diameter of the ducts increases as well, from 40 micrometers in the straight collecting tubules, to 100-200 micrometers in the papillary ducts.

Along the entire extent of the collecting duct, the principal cells stain only weakly with the usual stains. They have electron-lucent cytoplasm with few organelles and almost no invaginations of the basal cell membrane. They have a single cilium and relatively few, short microvilli (not visible with light microscope). They do not have processes that interdigitate with those of neighbouring cells. As a result, the cell boundaries are often quite evident, which is not the case in the other thick tubules. This is a very useful criterion when trying to distinguish collecting ducts from distal straight segments in areas where they are found together (medullary rays, outer medulla), especially in cases where the staining of the two types of tubules is not markedly different. (The brush border of proximal straight tubules makes them easy to distinguish, and they also stain more darkly.) In the inner medulla, it is easy to distinguish the tall cuboidal or columnar collecting ducts from thin segments of the loop of Henle.

Several other criteria can sometimes be useful in distinguishing collecting ducts from distal tubule (DT) straight segments. In DTs, nuclei tend to be located closer to the lumen due to the presence of basal mitochondria, whereas the nuclei of collecting ducts tend to be more basally located. The cells of the (early) collecting ducts tend to be smaller and more cuboidal than those of the DT. However, the outer diameter of collecting tubules is larger than that of DTs. The luminal spaces of collecting ducts tend to be more regular than those of DTs.

In addition to the principal cells described above, another type of cell, the dark cell or intercalated cell, can sometimes be seen. Dark cells are found in much smaller numbers, and their frequency decreases as the duct progresses. None are found near the papilla. Microplicae (cytoplasmic folds) and microvilli are present on their apical surface, and the apical cytoplasm has numerous vesicles. These cells do not have basal infoldings but do have basally located interdigitations with neighbouring cells. The function of dark cells is not known.

The permeability of the collecting ducts is regulated by antidiuretic hormone (ADH) from the posterior pituitary. In the absence of ADH, the principal cells are impermeable to water, and a dilute copious urine is produced. ADH promotes the permeability of the cells to water, which leaves the duct and enters the interstitial space. As a result, a small volume of hypertonic urine is produced. Excess water consumption inhibits the release of ADH. Inadequate consumption of water, or loss of fluid through sweating, vomiting or diarrhea stimulates the release of ADH. The inability to produce an adequate amount of ADH leads to diabetes insipidus.

Making Urine Hypertonic: The Countercurrent Multiplier Mechanism

All of the nephrons are involved in the filtering of blood, and the resoption and secretion of specific substances in the various segments of their tubules. Only the juxtamedullary nephrons with long loops of Henle are involved in the creation of a hypertonic urine. The manner in which this occurs is called the countercurrent multiplier mechanism. The long loops of Henle create a gradient of hypertonicity in the medullary interstitium that influences the concentration of the urine as it flows through the collecting ducts. The osmolarity of the interstitium increases from 300 mosm/L at the outer medulla adjacent to the cortex, to 1200 mosm/L in the deepest part of the inner medulla.

Recall that the thick descending (proximal) limb gives way to the thin descending limb near the outer part of the medulla. It is the thin descending limb and the entire ascending limb (thin and thick) that are involved in the countercurrent mechanism. The thin descending limb is freely permeable to Na, Cl and water, whereas both the thin and thick segments of the ascending limb are impermeable to water. Furthermore, the ascending limb actively transports Cl from the lumen to the interstitium, and Na passively follows to maintain electrochemical neutrality. Because water can't leave the ascending limb, the interstitium becomes hypertonic relative to the luminal contents. (Although some Na and Cl diffuse back into the nephron at the descending limb, these ions are transported out again at the ascending limb.) This produces the countercurrent multiplier effect, in which the concentration of NaCl increases in the interstitium down the length of the loop of Henle. The modified filtrate in the ascending limb that ultimately reaches the DCT becomes hypotonic.

In the absence of ADH, the DCT and the collecting tubules and ducts are also impermeable to water and a great quantity of hypotonic urine is formed. When ADH is present, they become highly permeable to water. Within the cortex, in which the interstitium is isotonic with blood, the modified filtrate equilibrates to isotonicity in the DCT, partly by loss of water to the interstitium and partly by addition of ions other than Na and Cl to the filtrate. In the medulla, increasing amounts of water leave the filtrate as the collecting duct progresses through the increasingly hypertonic interstitium. The filtrate (urine) becomes hypertonic.

The structure of the vasa recta that are closely associated with the loops of Henle is described above (under Blood Supply). In the medulla, these vessels form loops that closely parallel the loops of Henle. Because of this arrangement, these vessels can provide circulation to the medulla without disturbing the osmotic gradient that is established by the chloride pump in the ascending limb. Recall that both the arterial and venous limbs are very thin-walled and permeable. As the arterial vessels descend through the medulla, the blood loses water to the interstitium and gains salt from it, so that at the tip of the loop deep in the medulla, the blood is essentially in equilibrium with the hypertonic interstitial fluid. As the venous vessels ascend, the process is reversed, the hypertonic blood loses salt to, and gains water from, the interstitium. This passive countercurrent exchange of water and salt between the blood and interstitium occurs without the expenditure of energy. It is ultimately driven by the energy that drives the multiplier system: the active transport of Cl by the ascending limb of the loop of Henle.

The Excretory Passages

The urine that is excreted at the area cribrosa flows sequentially to: a minor calyx belonging to that papilla, a major calyx, the renal pelvis, through the ureter to the urinary bladder where it is stored, and through the urethra where it is voided.

All these passageways, except the urethra, have the same general structure:

- *a mucosa.* The mucosa consists of a special type of epithelium called transitional epithelium and an underlying lamina propria.
- *a muscularis.* The thickness and orientation of the muscle layers varies.
- an adventitia, or in some regions, a serosa.

There is no distinct submucosa. The lamina propria blends with the underlying smooth muscle of the muscularis.

Transitional Epithelium

The transitional epithelium that lines the excretory passages of the urinary system (except for the urethra) is unique to those regions. It is much thicker in the bladder and ureters (up to about 8 cells thick) than in the calyces (2 to 3 cell thick). It is essentially impermeable to salts and water.

It is also able to become thinner and flatter, thereby allowing distensibility of the passageways. When the urinary bladder is empty (undistended), its transitional epithelium is about 5-8 cells in thickness. When it is full (distended), its epithelium is about 3-4 cells in thickness. The ability of transitional epithelium to distend is due to the structure of the surface cells.

In routine histological sections of transitional epithelium (where the tissue is not distended), the surface cells appear large and rounded, and bulge into the lumen. Many are binucleate.

These large round surface cells become more or less squamous in the distended state. The basal cells of transitional epithelium are the smallest, and the nuclei appear crowded because of the minimal amount of cytoplasm in the cell. The cells in the intermediate layers are intermediate in size. Examination of the surface cells with the transmission electron microscope shows thickened, more rigid areas of the plasma membrane, called plaques, on the luminal surface. Between the plaques are shorter areas where plasma membrane is not thickened- the interplaque regions. In the undistended state, the plaques fold inward into the apical cytoplasm. In sections, they appear as fusiform vesicles because their continuity with the apical surface is not evident. In the distended state, the clefts and fusiform vesicles are reinserted into the cell surface as the mucosa unfolds and the cells flatten. Filaments attached to the undersurface of the plaque and extending into the apical cytoplasm may prevent undue stretching in the distended state. The luminal membrane of surface cells has an unusual chemical compostion: cerebroside is the major component of the polar lipid fraction.

The Ureters

The ureters convey the final urine to the bladder. They arise from the renal pelvis and follow an oblique path through the wall of the bladder. In histological sections, the mucosa is typically folded, having a star-shaped appearance, due to the contraction of the smooth muscle of the muscularis. The lamina propria is fairly wide, and consists of fibroelastic connective tissue, usually more dense and with more fibroblasts under the epithelium, and looser near the muscularis. Diffuse lymphatic tissue may be seen. The muscularis consists of an inner longitudinal and an outer circular layer of smooth muscle. They are usually not clearly defined as two layers. They generally appear as loose, anastomosing strands of smooth muscle separated by abundant collagenous connective tissue. In the terminal part of the ureters, a thick outer layer of longitudinal muscle is described in addition to the other two, particularly in the portion of the ureter that passes into the bladder. Peristaltic contractions of the smooth muscle move the urine from the minor calyces through the ureter to the bladder.

As the bladder distends with urine, the openings of the ureters are compressed, reducing the possibility of reflux of urine into the ureters. Contraction of the smooth muscle of the bladder wall also compresses these openings. This is important in protecting the kidney from the spread of infections from the bladder and urethra, which are frequent sites of chronic infection, especially in the female.

The ureters are covered with an adventitia which is continuous with the surrounding fibroelastic connective tissue containing abundant adipose cells, numerous blood vessels and small nerves. A portion of the tube may be covered with a serosa.

The Bladder

Urine is temporarily stored in the bladder before it is eliminated to the outside. The size and shape of the bladder change as it fills. The bladder has three openings, two for the ureters and one for the urethra. The triangular region formed by these openings, called the trigone, is relatively smooth and constant in thickness. The rest of the bladder wall is thick and folded when the bladder is empty, and thin and smooth when it is full. The differences between the trigone and the rest of the bladder wall reflect their different embryological origins: the trigone originates from the embryonic mesonephric ducts, while the major portion of the wall originates from the cloaca.

The layers of the bladder wall are similar to those of the ureter. The mucosa looks similar to that of the ureter, but its thick muscularis is distinctive. The muscularis has three layers of muscle, an inner longitudinal, a middle circular or spiral layer and an outer longitudinal. The middle layer is the largest. However, the muscle layers are even less clearly defined than in the ureter, and appear as anastomosing bundles separated by loose connective tissue. In the region of the trigone, thin, dense bundles of smooth muscle form a circular mass around the internal opening of the urethra, forming the internal sphincter of the bladder. Contraction of the smooth muscle of the bladder forces the urine into the urethra. The bladder has a serosa, covered with mesothelium, on its upper surface. Elsewhere, the outer layer of the wall consists of a fibrous adventitia.

The Urethra

The urethra is the terminal part of the excretory passageway. In the male, it also serves to convey sperm to the outside.

Male urethra: The male urethra is about 20 cm long and can be divided into three distinct segments. The first segment, the prostatic urethra, extends from the neck of the bladder through the prostate gland. The next segment, the membranous urethra, extends for about 1 cm from the apex of the prostate gland to the bulb of the corpus spongiosum of the penis. The last segment, the penile urethra, extends for about 15 cm through the length of the corpus spongiosum and opens to the outside at the glans. A loose, fibroelastic lamina propria underlies the epithelium of the urethra (described below for the various segments). The prostatic urethra is lined by transitional epithelium. It is surrounded by the fibromuscular stroma of the prostate gland. It lies on a conical elevation of particularly dense stroma without glands called the colliculus seminalis. The colliculus protrudes into the lumen of the urethra and gives it a crescent shape. The prostatic utriculus, a remnant of the embryonic Mullerian duct, lies in the mass of the colliculus and opens into the urethra. Ejaculatory ducts (the continuations of the vasa deferentia into the prostate) course alongside the utriculus and finally open into the urethra. In addition, there are numerous small openings of the ducts of the prostatic glands into the urethra. (The prostatic glands are described in the Reproduction Block. The sections of the prostate on our slides do not contain the urethra.). The membranous urethra is lined with pseudostratified or stratified columnar epithelium that more closely resembles the epithelium of the genital duct system. It is surrounded by the skeletal muscles of the pelvic and urogenital diaphragms, which form the external (voluntary) sphincter of the urethra. The penile urethra, like the membranous part, is lined with pseudostratified or stratified columnar epithelium, but it can have patches of stratified squamous epithelium. Where the urethra opens at the tip of the glans, the lumen becomes dilated and star-shaped, and is called the fossa navicularis. The fossa navicularis has stratified squamous epithelium which is continuous with epidermis of the skin.

The surface of the mucosa of the urethra has many recesses (lacunae of Morgagni) which continue into deeper, branching mucous glands, the glands of Littre. These are best developed in the penile urethra. Mucous cells, singly or in groups, are also found along the surface epithelium.

The penile urethra is surrounded by the erectile tissue of the corpus spongiosum, which contains numerous large, irregular venous spaces separated from each other by trabeculae of fibroelastic connective tissue with numerous smooth muscle fibres.

Female urethra: The female urethra is short, 3-5 cm in length, extending from the bladder to the vestibule of the vagina, where it terminates just below the clitoris. The mucosa has many longitudinal folds. Near the bladder, the epithelium may be transitional, but for most of the length of the urethra it is stratified squamous. Patches of stratified columnar and pseudostratified columnar epithelium may be seen. Mucus-secreting cells in the mucosal folds, similar to the glands of Littre, are present. Numerous small glands, the para- or periurethral glands, which are homologous to the prostate in the male, open into the urethra. The lamina propria is a highly vascularized layer of connective tissue with many venous sinuses; it resembles the corpus spongiosum. The muscularis has two poorly-defined layers: inner longitudinal and outer circular. Where the urethra penetrates the urogenital diaphragm, the striated muscle of this structure forms the external or voluntary urethral sphincter. The outer adventitia blends with that of the vagina.

Male Reproductive System

The male reproductive system consists of the testes, ductuli efferentes, epididymides, vasa deferentia, seminal vesicles, prostate gland, bulbourethral glands and penis.

Low Power View of Testis

Male gametes (sperm) and male sex hormones (mainly testosterone) are produced in the testis. Numerous profiles of seminiferous tubules are seen, most of them cut more or less in cross section. The epithelium of the seminiferous tubules is an indistinct stratified cuboidal, and is composed of the gametogenic cells in various stages of development. The earliest stages (spermatogonia) are found closest to the periphery, the later stages are found progressively toward the lumen. Sperm (also called spermatozoa) can sometimes be seen along the luminal part of the epithelium or in the centre of the lumen. (Don't try to identify cell types at this magnification). The boundary of each seminiferous tubule profile is fairly well demarkated. This reddish border consists of the basement membrane and the surrounding connective tissue (considered as the lamina propria; in humans it has contractile myoid cells rather than typical fibroblasts). Between the seminiferous tubules is a loose connective tissue. In this interstitial tissue are found the interstitial cells of Leydig (or Leydig cells), which produce testosterone. (Don't try to find them at this magnification). The testis is surrounded by a thick fibrous capsule called the tunica albiguinea.

High Power View of a Seminiferous Tubule

Spermatogonia are found right along the basement membrane. Spermatogonia are the stem cells. They can divide by mitosis to produce more spermatogonia (type A spermatogonia) or they can become committed to differentiating into primary spermatocytes (type B spermatogonia). Type A spermatogonia stain more darkly and have an ovoid nucleus with finely granular chromatin. Type B are paler-staining, generally with more spherical nuclei and more condensed chromatin. (You are not required to distinguish the different types of spermatogonia.)

Committed spermatogonia will differentiate to become a primary spermatocyte. Primary spermatocytes undergo the first meiotic division.

They are the largest cells of the spermatogenic series, with large nuclei exhibiting various stages of meiosis I. They lie more centrally than the spermatogonia.

When a primary spermatocyte completes the first meiotic division, it gives rise to two secondary spermatocytes. Secondary spermatocytes complete the second meiotic division, each one giving rise to two spermatids. However, they are almost never seen. Whereas primary spermatocytes take about three weeks to complete meiosis I, secondary spermatocytes complete meiosis II within a few hours. Therefore generally only spermatids are seen on the luminal side of primary spermatocytes.

Spermatids are much smaller than primary spermatocytes. Spermatids are genetically complete, that is, they contain the exact genetic information that the sperm will carry. However, they undergo a great morphological change, during a process called spermiogenesis, in which they change into motile, flagellated cells. Early spermatids are round, whereas later stages look more and more like sperm.

Spermiogenesis

Spermiogenesis is the process undergone by spermatids to become motile, flagellated cells. In the first stage (Golgi phase), pro-acrosomal granules begin to accumulate in the Golgi complex and coalesce within a membrane-lined acrosomal vesicle adjacent to the nuclear envelope. The location of this vesicle determines the anterior pole of the sperm. The centrioles migrate to the posterior pole, and the distal centriole aligns itself at right angles to the plasma membrane. The distal centriole initiates the formation of the flagellar axoneme.

The acrosomal vesicle then spreads out to cover the anterior half of the nucleus (cap phase). The acrosomal vesicle and nucleus become further condensed, and the nuclear envelope thickens and loses its pores. The acrosome contains enzymes, including hyaluronidase, neuramidase, acid phosphatase and a trypsin-like protease.

The spermatid then reorients itself to face toward the base of the seminiferous tubules, with its head deeply embedded in a Sertoli cell (acrosome phase). The nucleus becomes elongated and condensed; it and the acrosome become closely apposed and lie immediately adjacent to the anterior plasma membrane. The acrosome and nucleus will form the head of the sperm. The cytoplasm is deisplaced posteriorly. The cytoplasmic microtubules become organized into a cylindrical sheath, the manchette, that extends from the posterior rim of the acrosome to the posterior pole of the spermatid.

The centrioles move back to the posterior surface of the nucleus and become modified to form a connecting piece or neck. Nine coarse fibres develop from the cenrioles attached to the nucleus and extend into the tail as the outer dense fibres peripheral to the microtubules of the axoneme. The plasma membrane moves posteriorly to cover the growing flagellum. The manchette disappears. The mitochondria aggregate around the proximal part of the flagellum as a helically-wrapped sheath, forming a thickened region called the middle piece. Distal to this, a fibrous sheath surrounds the axoneme and extends nearly to the end of the flagelllum, forming the principal piece. During the last stage, the maturation stage, excess cytoplasm is pinched off as the residual body which is phagocytized by Sertoli cells. The intercellular bridges remain attached to the residual bodies, and the sperm become independent of one another.

Sertoli Cells

Sertoli cells are found in the seminiferous tubules but are not part of the spermatogenic series. They are sometimes called "nurse" cells because of their role in serving the spermatogenic cells. Sertoli cells are long, pyramidal cells that sit on the basement membrane and can extend all the way to the lumen. Their many processes surround the spermatogenic cells. However, usually only their nuclei are visible in the light microscope (sometimes – rarely – you can see a bit of cytoplasm). Their nuclei are pale, elongated, often triangular, and have a dark, prominent nucleolus. Often you will see a number of Sertoli cell nuclei in a row.

The major roles of Sertoli cells are (1) the support, protection and nutrition of spermatogenic cells, (2) phagocytosis and (3) secretion. The cytoplasmic ramifications of Sertoli cells provide physical support to the network of cytoplasmic bridges of the spermatogenic cells.

Sertoli cells form the "blood-testis barrier" that divides the seminiferous tubules into a basal compartment (with spermatogonia) and an adluminal compartment (other cells of spermatogenic series). The basal compartment has free access to the blood; the adluminal compartment, and the rest of the excurrent ducts, are protected from blood-borne products and differ from blood plasma in their ionic, amino acid, protein and carbohydrate compositon.

This arrangement protects developing sperm from immunogenic attack. The haploid germ cells develop after the individual becomes immunologically competent and would therefore be recognized as foreign by the immune system. Adjacent Sertoli cells are bound together by occluding junctions, gap junctions and cytoplasmic specializations involving sER cisternae and actin filament bundles. To let early spermatocytes past this barrier to reach the adluminal compartment, Sertoli cells form new junctional complexes below newly-formed spermatocytes, while the ones above break down. The cells in the adluminal compartment depend on Sertoli cells for the exchange of nutrients and metabolites.

The main phagocytic function of Sertoli cells is the removal of residual bodies. They secrete a fluid that facilitates the passage of sperm into the genital ducts. They also secrete an androgen-binding protein (ABP) that concentrates testosterone in the tubule lumen; high levels are needed for the normal development of sperm. The also secrete inhibin to regulate spermatogenesis. Inhibin inhibits the synthesis and release of follicle-stimulating hormone from the anterior pituitary.

Leydig Cells

Leydig cells are the steroidogenic cells of the testis. They are found singly or in clusters in the connective tissue between the seminiferous tubules. They are large cells with extensive eosinic cytoplasm and variable numbers of lipid vacuoles. Ultrastructurally they are typical steroid-secreting cells. Their major product is testosterone.

Prostate Gland with Concretions

The prostate gland produces much of the seminal fluid, and includes in its secretions acid phosphatase, fibrinolysin and citric acid. Pink-staining secretions are frequently seen in the lumina of the glands. Prostatic concretions, also called corpora amylacea, are frequently found in the glands of older men. They result from the precipitation of secretory material around cell fragments, and can become calcified. The penis is the terminal part of the reproductive system and the urinary system. It consiststs of two dorsal erectile bodies, called corpora cavernosa, and one smaller, ventral erectile body, called the corpus spongiosum (or corpus cavernosum urethrae). The urethra runs through the corpus spongiosum. The three erectile bodies are held together by a connective tissue sheath called the tunica albiguinea, which forms a capsule around each one. (The capsules between the two corpora cavernosa meet in the midline to form the medial septum, which becomes attenuated near the tip of the penis.) The capsules around the two corpora cavernosa are very dense, and cause them to become very rigid when they fill with blood during an erection. The capsule around the corpus spongiosum is thinner and more elastic, so it does not become as rigid as the other two bodies. This prevents the urethra from becoming occluded during an erection. The corpus spongiosum expands at the tip of the penis to form the glans. The corpora cavernosa end at the base of the glans. Surrounding the three encapsulated erectile bodies is very loose connective tissue, called Buck's fascia or deep penile fascia, which in turn is surrounded by the dermis and epidermis of thin skin. Because of the loose CT, the skin moves easily over the underlying stuctures.

In the specimens in our collection, the sections of the penis were made close to the base, when the crura bearing the corpora cavernosa, and the root bearing the corpus spongiosum, are still approaching each other. Thus, these structures are much farther apart (separated by more connective tissue), than in the sections typically shown in histology textbooks. Even at low power, it is not be possible to have more than one body (corpus cavernosum or corpus spongiosum) in the field of view.

Embryology of the Reproductive System

The reproductive system arises in close association with the urinary system, and the two associated systems are often called the urogenital system. The urogenital system arises from the intermediate mesoderm. The urinary system arises before gonadal development begins. The early development of the genital system is the same in both sexes and is called the undifferentiated stage.

Kidneys and Internal Genital Organs

Develoment of the urinary system begins on about day 22, when 5 to 7 paired cervical segments give rise to nephric vesicles or nephrotomes. In humans these are vestigial and never differentiate into a functional state; in some lower vertebrates they would develop into functional embryonic pronephroi.

Early in the fourth week, nephric tubules begin to develop within a pair of elongated swellings-the mesonephric ridges-from the upper thoracic to the third lumbar level. About 40 mesonephric tubules are produced in craniocaudal succession. The cranial ones regress as the more caudal ones form, so there are never more than 30 pairs. By the fifth week, there is a massive regression of the cranial ones, leaving about 20 pairs in the first 3 lumbar levels.

These mesonephric tubules will develop into excretory units (renal corpuscle plus tubule) that look like an abbreviated version of the adult (metanephric) kidney tubules. Their duct, the bilateral mesonephric or Wolffian duct, arises on about day 24 as a pair of solid longitudinal rods. The rods grow caudally towards and fuse with the ventrolateral surface of the cloaca (about day 26). The region of fusion will become part of the posterior wall of the future bladder. As the rods fuse with the cloaca, they begin to canalize (develop a lumen) from their caudal toward their cranial end.

The mesonephric tubules fuse with the mesonephric duct which provides a passage to the cloaca. The mesonephric excretory units function from about weeks 6 to 10, producing a small amount of urine. After week 10, they stop functioning and regress.

NB: In the male, a few modified mesonephric tubules and the mesonephric duct will persist. The former will become the ductuli efferentes (efferent ductules) of the testis and the latter will giver rise to the epidydimis and vas deferens. In the female, these structures will regress. The formation of the definitive kidneys (metanephric kidneys) begins on about day 28. Ureteric buds (which give rise to the future ureters and collecting duct system) sprout from the distal part of the mesonephric ducts. The buds induce the intermediate mesoderm of the sacral region to form the metanephric blastema (future nephric tubules). There are reciprocal inductions between the branching ureteric bud and the metanephrogenic mesoderm in the development of the kidneys. (The development of the kidney will not be discussed further.) Probably due to differential growth of the lumbar and sacral regions, the kidneys ascend to just below the adrenal glands between the sixth and ninth week.

Between the fourth and sixth week, the cloaca septates to give rise to 2 structures: the rectum posteriorly and the urogenital sinus anteriorly. The urogenital sinus gives rise to: 1) the bladder in both sexes; 2) the membranous (= entire) urethra in females, and the membranous and prostatic urethra in males; 3) the vestibule of the vagina in females and the penile urethra in males.

While the urogenital sinus is forming, the mouths of the mesonephric ducts expand and flatten and blend into the wall of the part that will become the bladder. The ureteric buds (future ureters) arising from the mesonephric ducts intercalate into the future bladder's posterior wall. When the mesonephric ducts regress in females, the ureters persist.

During the sixth week, a new pair of ducts, the paramesonephric or Müllerian ducts, begin to form just lateral to the mesonephric ducts. They arise by a craniocaudal invagination of a ribbon of thickened coelomic epithelium from the third thoracic segment to the posterior wall of the urogenital sinus. Their caudal tips grow to connect with the urogenital sinus just medial to the openings of the two mesonephric ducts. The tips adhere to each other just before contacting the urogenital sinus.

In female embryos, as the Müllerian ducts contact the urogenital sinus, they begin to fuse from their caudal tips cranially (between third and fifth month), forming a tube with a single lumen, the uterovaginal canal (future uterus and upper vagina). The unfused portions of the Müllerian ducts give rise to the Fallopian tubes. The posterior urogenital sinus contacted by the Müllerian ducts thickens to form a pair of swellings called sinovaginal bulbs, which will give rise to the lower fifth of the vagina.

NB: In males, the paramesonephric ducts will regress under the influence of Müllerian inhibiting hormone (MIH) produced by pre-Sertoli cells of the developing testis, which normally begins to differentiate at the beginning of the seventh week.

The Development of the Gonads

The initial stage of gonadal development occurs during the fifth week when a thickened area of mesodermal epithelium develops on the medial side of the mesonephros. Proliferation of the epithelium and the underlying mesenchyme produces a bulge called the gonadal or genital ridge. Fingerlike epithelial cords, called primary or medullary sex cords, grow into the ridge. The genital ridge is also invaded by primordial germ cells (future spermatogonia or oogonia) from the yolk sac. During folding of the embryo, the dorsal part of the yolk sac is incorporated into the embryo. The primordial germ cells migrate along the dorsal mesentary of the hindgut to the gonadal ridges. During the sixth week, they enter the underlying mesenchume and begin to be incorporated into the sex cords. Up to the beginning of the seventh week, male and female embryos appear morphologically identical. In both sexes, germ cells and sex cords are present in the cortical and medullary regions of the undifferentiated gonad, and complete mesonephric and paramesonephric ducts lie side by side.

At the beginning of the seventh week, the sex cord cells of male embryos begin to produce testis determining factor (TDF). The gene for TDF is located on the short arm of the Y chromosome. This region is called the sex-determining region of the Y chromosome (SRY). Under the influence of TDF, cells in the medullary region of the primitive sex cords differentiate into (pre-)Sertoli cells, while those in the cortical region degenerate.

The differentiating pre-Sertoli cells organize to form testis cords. At puberty, the testis cords will canalize to form the seminiferous tubules. Testis cords that are distal to the germ cell region will give rise to the rete testis. During the seventh week, the testis also rounds up, reducing its contact with the mesonephros, which exerts a feminizing influence. The cell-to-cell contact between pre-Sertoli cells and the primordial germ cells appears to play a role in the development of male gametes. Pre-Sertoli cells secrete Müllerian inhibiting hormone (MIH) (also called Müllerian inhibiting substance or anti-Müllerian hormone). MIH (which closely resembles transforming growth factor Beta) causes the paramesonephric ducts to regress. It is also the probable inducer for the differentiation of the androgen-producing Leydig cells in the mesenchyme of the genital ridges. The andorgens are necessary for the mesonephric duct to give rise to the male genital ducts. Pre-Sertoli cells also inhibit germ cell development before meiosis begins. In female embryos, TDF is not produced as no Y chromosome is present. Development is slowed. The primordial germ cells remain concentrated in the outer cortical region or near the cortico-medullary boundary. Pre-Sertoli cells do not arise from the primary sex cords, which form a rete in the medulla and degenerate (tenth to twelfth week). Secondary (or medullary) sex cords invaginate from the germinal epithelium and incorporate the primordial germ cells.

The follicle cells of the ovary arise from the secondary sex cords, or possibly from both the secondary and some primary sex cord cells. The female germ cells differentiate into oogonia (apparently with some influence from the mesonephros) and enter meiosis to become primary oocytes, before further differentiation is inhibited by the follicle cells. Individual oocytes resume gametogenesis at puberty.

The number of primordial follicles is maximum at 5 months (about 7 million). At birth, there are about 2 million, and by puberty only 40,000 are left. About 400 will become secondary oocytes.

The Development of the External Genitalia

The external genitalia develop from the same primordia in both sexes. Early in the fourth week, proliferating mesenchyme produces a genital tubercle at the cranial end of the cloacal membrane. The genital tubercle soon elongates to become a phallus. At the same time, labioscrotal folds (or genital folds) and urogenital (or urethral) folds develop on each side of the cloacal membrane. While the development of male internal genital ducts depends on testosterone, the development of male external genitalia depends on the metabolite of testosterone, dihydrotestosterone (DHT). DHT causes the phallus to enlarge and elongate and form a penis. As this occurs, the urogenital folds develop and form the lateral walls of the urethral groove of the ventral surface of the penis. The labioscrotal folds fuse to form the scrotum.

In females, the pattern of the external genitalia more closely resembles the undifferentiated stage. The phallus regresses to become the clitoris, the urogenital folds become the labia minora, and the labioscrotal folds become the labia majora. The urogenital sinus remains open as the vestibule, into which both the urethra and the vagina open.

The Mammary Gland

Development of Mammary Glands

The mammary glands, or breasts, are branched tubulo-alveolar glands. They are believed to be modified apocrine sweat glands. They develop as a pair of epidermal thickenings, called milk lines or mammary ridges, in the four-week old embryo of both sexes. Multiple glands develop along the ridges which extend from the axilla to the inguinal region. In humans, only one pair usually persists, and the ridges disappear elsewhere.

In the fifth week, the primary bud of the gland grows down into the underlying dermis. It begins to branch in the tenth week. By the 12th week, secondary buds begin to form, and will lengthen and branch throughout gestation. In the second trimester fetus, 15-25 ingrowths of the epithelium penetrate the underlying connective tissue. They will give rise to the lactiferous ducts, through canalization of the buds of cells.

The lactiferous ducts open into a small superficial depression called the mammary pit. Proliferation of the underlying mesoderm converts the pit into an everted nipple within a few weeks after birth. The skin around the nipple also proliferates to form the areola.

In newborns of both sexes, the ducts may be swollen with secretory material due to the influence of placental and maternal hormones. The exudate sometimes seen on the nipples of newborns is called angel milk or witch's milk. In males, the mammary glands remain rudimentary. Experiments in mice have shown that testosterone causes an involution of the developing ducts and that its effect is mediated throught the mammary mesenchyme rather than acting directly on the ductal epithelium. In females before puberty, the glands are composed of the lactiferous ducts which converge at the nipple. The development of the breasts during puberty constitutes one of the female secondary sexual characteristics.

Sometimes supernumerary breasts (polymastia) or nipples (polythelia) develop along the milk lines. The latter are as common in males as in females. The most common location is just below the normal breast. More rarely, an ectopic nipple forms off the line of the mammary ridge as a result of migration of mammary tissue.

Structure of Mammary Gland in Mature Female

Each mammary gland has 15-25 lobes of branched, tubuloalveolar glands which radiate from the nipple (mammary papilla). The lactiferous duct is the single excretory duct for each lobe. Beneath the areola, each lactiferous duct dilates into a lactiferous sinus, then narrows again before exiting at the summit of the nipple. The lobes are separated by fibrous bands of connective tissue. Some of these bands connect with the dermis and are called the suspensory ligaments of Cooper. The lobes are further subdivided into lobules. A lobule consists of several intralobular ducts (that is ducts within the lobules), which empty into a terminal interlobular duct (duct between lobules). The interlobular ducts empty into one of the lactiferous ducts.

The extent of the development of the gland depends on whether it is in a non-lactating (inactive, resting) or lactating (active, secretory) state. The glandular elements of the adult, inactive mammary gland consist of a system of ducts from which large numbers of secretory elements develop during pregnancy.

Inactive mammary gland: In the inactive mammary gland, the glandular elements consist only of ducts. The intralobular ducts and the terminal interlobular ducts are lined with simple cuboidal, sometimes low columnar, epithelium, supported by a prominent basement membrane. A discontinuous layer of myoepithelial cells (or basket cells) lies between the intralobular duct cells and the basement membrane. These cells are epithelial in origin but show many features of smooth muscle cells. They have long, thin, cytoplasmic arms that extend around the duct cells in order to squeeze out the contents. The lactiferous ducts are lined by cuboidal or columnar epithelium for most of their length, by two layers of cuboidal cells in the lactiferous sinus, and, near their opening, by stratified squamous epithelium.

Within the lobules, the connective tissue that surrounds the intralobular ducts is loose and highly vascular. It contains a large number of cells such as fibroblasts, lymphocytes and plasma cells. The lobules have little or no fat.

The lobules are separated from one another by dense connective tissue that may contain abundant adipose tissue. In histological sections, they appear as lightly-staining islands in the surrounding dense connective tissue.

The inactive mammary gland undergoes slight changes during the menstrual cycle. Early in the cycle, the ducts appear as cords with little or no lumen. During the progestational (secretory) phase, the cells increase in height and very slight development of secretory alveoli may begin. Lumina appear in the ducts along with small amounts of secretion, and fluid accumulates in the connective tissue.

Active mammary gland: Dramatic changes in the mammary gland occur during pregnancy. The ducts in the lobules proliferate and branch and secretory alveoli sprout from them. Alveoli (or acini) are spherical collections of epithelial cells that become the active milk-secreting structures in lactation. The alveoli are made of cuboidal or low columnar epithelium with myoepithelial cells at their base. They are difficult to distinguish from the ducts (and you are not required to do so). The increase in glandular tissue is accompanied by a great decrease in the amount of interlying connective and adipose tissue. These changes progress during pregnancy. After birth, when the mammary gland is actively secreting, alveoli and their ducts completely predominate. Lobules are separated only by greatly reduced strands of dense interlobular connective tissue. Plasma cells, lymphocytes and eosinophils infiltrate the fibrous component of the connective tissue. The lumina of the alveoli also become much larger during pregnancy and especially after parturition, and secretory material is seen inside them. Technically, the term active mammary gland is reserved for the post-natal lactating mammary gland. During pregnancy, the gland is called proliferative.

Breast Proliferation and Lactation

Breast proliferation depends on estrogens (affect mainly ducts) and progesterone (stimulates growth of alveoli) from the corpus luteum and placenta, and also on prolactin, human placental lactogen (or human chorionic somatomammotropin), thyroid hormone and corticosteroids. As pregnancy progresses, the alveoli begin to secrete a protein-rich fluid called colostrum. It contains less fat, carbohydrate and potassium than milk, but more protein, vitamin A, sodium and chloride. It also contains antibodies which provide passive immunity to the baby. The antibodies are thought to be produced by the lymphocytes and plasma cells that infiltrate the stroma of the breast during its proliferation. They are secreted across the glandular cells. The number of these cells decreases after parturition. A few days after birth, colostrum is replaced by lipid-rich milk. The accumulation of secretory product dilates the alveolar and duct lumina and can be seen in histological sections.

Lactation

The lactating breast is composed almost entirely of alveoli distended with milk with thin septa of connective tissue between lobules. The secreting cells contain abundant granular endoplasmic reticulum, a moderate number of large mitochondria, a supranuclear Golgi complex and a number of dense lysosomes. Lipid droplets and secretory granules may be present in the apical cytoplasm.

The lipid component of the milk is released by apocrine secretion. It arises as lipid droplets free in the cytoplasm. The droplets coalesce to form large droplets that travel to the apical end of the cell. They become invested with an envelope of plasma membrane as they are released. A thin layer of cytoplasm is trapped between the plasma membrane and the lipid droplet. The protein component of the mile is released by merocrine secretion. The protein is synthesized in the rough ER, and packaged into membrane-lined secretory granules in the Golgi apparatus. It is released by fusion of the granule's limiting membrane with the plasma membrane of the cell. The lactating gland resembles the thyroid gland; however, all of the alveoli in a mammary gland are joined to a duct and the place where several alveoli open into a central channel can often be seen. Large ducts can also be seen in the connective tissue septa. (The thyroid, an endocrine gland, has no ducts).

Hormones of Lactation

A number of hormones are involved in the control of lactation. The proliferation of the duct system in early pregnancy is stimulated by estrogens, glucocorticoids and growth hormone. Progesterone and prolactin are required for the final steps of lobular maturation, and in humans, placental lactogen (also called chorionic somatomammotrophin) may alse be required. Prolactin is involved in the production of milk sugars, fatty acid synthesis and production of milk fats, and elctrolyte metabolism (Na/K pump). Estradiol stimulates prolactin secretion and increases the number of mammary gland prolactin receptors, while progesterone decreases the number of receptors. Prolactin also upregulates its own receptors.

Placental lactogen has effects similar to those of growth hormone in promoting the growth, differentiation and protein synthesis in mammary glands. Both are involved in lipid mobilization. The glucocorticoids and insulin are required to maintain viability, growth and differentiation of mammary gland cells. The glucocorticoids are involved in water and electrolyte metabolism. Insulin promotes glucose uptake in the active mammary gland, and participates in stimulating the production of milk proteins and lipids.

The major hormones involved in lactation are summarized below:

Estrogen: Stimulates development of duct system.

Progesterone: Stimulates development of alveoli.

Prolactin: Stimulates alveolar maturation, involved in milk sugar and protein production, and water and electrolyte metabolism.

Human placental lactogen (h. chorionic somatomammotropin): Similar to growth hormone. Stimulates growth, differentiation, and protein synthesis in mammary gland. Synergizes with growth hormone in lipid mobilization.

Corticosteroids: Maintains viability of, and involved in growth and differentiation of mammary gland cells. Involved in water and electrolyte metabolism.

Insulin: Stimulates production of milk proteins and lipids. In the active mammary gland, promotes glucose uptake.

Thyroid hormones: General regulation of metabolic processes.

Relaxin: A hormone produced by decidual cells that stimulates the softening of the cervix and pelvic ligaments in preparation for childbirth.

Oxytocin: Stimulates contraction of myoepithelial cells for milk release.

Suckling

The act of suckling initiates sensory impulses in the nipple that are conveyed to the hypothalamus. The impulses inhibit prolactin-inhibiting factor, and prolactin is released from the adenohypophysis. The impusles also cause the release of oxytocin from the neurohypophysis. Oxytocin stimulates the contraction of the myoepithelial cells surrounding the base of the alveoli and ducts causing milk ejection. In the absence of suckling, the secretion of milk ceases. When lactation ceases, the mammary gland reverts to the inactive state. The alveoli eventually undergo complete degeneration and fat and connective tissue increase.

Involution

After menopause, the duct system gradually atrophies, with only a few scattered ducts remaining- a condition similar to that of the prepubertal gland. The connective tissue becomes increasingly dense and homogeneous.

Histology of Cartilage and Bone

Cartilage and bone are specialized connective tissues. (It may be useful to review the connective tissue section of Fundamentals of Histology at this point). Like all connective tissues, cartilage and bone consist of cells and extracellular matrix.

The matrix of all CTs consists of fibers (collagen, reticular, and elastic) and amorphous ground substance, which contains proteoglycans and hyaluronic acid. Different connective tissues vary in the abundance and types of cells, in the proportion of cells to matrix, and in the properties of the matrix itself, including the numbers and types of fibres and the nature of the ground substance. The matrix is secreted by some of the cells in connective tissues, fibroblasts in most CTs. In cartilage, it is chondroblasts and chondrocytes that produce the matrix, while in bone, it is osteoblasts and osteocytes.

Cartilage: Cartilage is a solid connective tissue that is to a certain extent pliable, making it resilient. These characteristics of cartilage are due to the nature of its matrix. The ground substance of cartilage is rich in proteoglycans consisting of a core protein with numerous- about 100- glycosaminoglycans (GAGs) attached bottle-brush fashion around it. GAGs are made of repeating units of disaccharides, one of which is always a glycosamine (hence the name) such as glucosamine or galactosamine. (Glycosamines are also called hexosamines.) In cartilage, the GAGs attached to the core proteins are chondroitin sulfate and keratan sulfate.

The proteoglycans themselves are attached, by special linker proteins to long, rigid molecules of hyaluronic acid (HA). HA itself is a GAG, but is composed of several thousand disaccharide units, rather than several hundred or less, as are other GAGs. About eighty proteoglycans are attached to one molecule of HA.

[For those who thirst for knowledge: The repeating units of chondroitin sufate are D-glucoronic acid and N-acetylgalactosamine-(4 or 6)-sulfate. The repeating units of keratan sulfate are galactose or galactose 6-sulfate and N-acetylglucosamine 6-sulfate. The repeating units of hyaluronic acid are D-glucuronic acid and N-acetylglucosamine].

In addition to the giant HA-proteoglycan macromolecules, there are other proteoglycans as well as glycoproteins in the matrix. The matrix also has collagen fibers, but these are of a finer nature (collagen Type II vs. Collagen Type I) than the collagen fibers in most other CTs. The macromolecules are bound to the thin collagen fibres by electrostatic interactions and cross-linking glycoproteins.

Between 60 and 80 percent of the net weight of (hyaline) cartilage is water, and this large component of water accounts for the resilient nature of cartilage. Water is attracted to the negative charges in the abundant sulfate and carboxyl groups on the GAGS. This hydration permits diffusion of water-soluble molecules in the ground substance. However the movement of large molecules and bacteria is inhibited. Cartilage is poorly vascularized, and gets most of its nutrients through diffusion. In the adult, repair is poor. The sulfate groups of the GAGs that attract water also attract molecules of basic dye. Differences in the intensity of staining within the matrix reflect differences in the abundance of proteoglycans. Staining is most intense around the cartilage cells. The collagen fibres are not visible in standard histological preparations as their refractive index is not markedly different from that of the ground substance.

Some basic dyes used to stain cartilage can undergo a colour change called metachromasia. In the presence of the negatively charged groups in the GAGs, the dye molecules form aggregates of dimers or polymers which changes their absorptive properties. Rather than staining blue, the tissue will stain purple or reddish.

Their are three kinds of cartilage, hyaline cartilage, elastic cartilage and fibrocartilage.

Hyaline Cartilage: Hyaline cartilage is the most abundant type of cartilage. Most of the skeleton of the fetus is laid down in cartilage before being replaced by bone. Hyaline cartilage in the adult is found in the nose, parts of the respiratory tract, at the ends of ribs and at the articular surfaces of bones.

Elastic Cartilage: The structure of elastic cartilage is very similar to that of hyaline cartilage, but in addition to the other components, its matrix has elastic fibres and interconnecting sheets of elastic material. This gives elastic cartilage an elasticity not present in hyaline cartilage. Elastic cartilage is found in the external ear, the walls of the external auditory canal, the Eustachian tube, the epiglottis and the larynx.

Fibrocartilage: Fibrocartilage has characteristics intermediate between those of hyaline cartilage and dense connective tissue. Its presence indicates the need for resistance to compression and shear forces. It is found in the intervertebral disks, the symphysis pubis, the articular discs of the sternoclavicular and temperomandibular joints, the menisci of the knee joint and some places where ligaments or tendons attach to bones. The amount of cartilage in fibrocartilage is variable, it generally occupies a smaller amount of the tissue. There is no perichondrium.

Embryology of Cartilage: Cartilage arises from mesenchyme. Some mesenchyme cells aggregate to form a blastema. The cells of the blastema begin to secrete cartilage matrix and are then called chondroblasts. They move apart as they deposit matrix, and when they are completely surrounded by matrix they are called chondrocytes. The mesenchymal tissue surrounding the blastema gives rise to the perichondrium.

Bone: Bone is a connective tissue distinguished by the fact that its matrix is mineralized by calcium phosphate in the form of crystals very similar to hydroxyapatite. The minerals are both in the (Type I) collagen fibers which constitute about 95% of the matrix, and in the ground substance. Bone is both resilient and hard. Its resilience is due to the organic matter (collagen), its hardness due to the inorganic minerals. Bone serves as a storage site for calcium and phosphate. Blood calcium levels are regulated by the hormones parathormone (parathyroid hormone), which raises blood calcium levels by stimulating bone resorption, and calcitonin, which reduces blood calcium by suppressing bone resorption and increasing osteoid calcification. (Osteoid is the matrix secreted by osteoblasts and osteocytes prior to mineralization). There are two kinds of mature bone: compact bone and spongy bone. Compact bone is also called dense bone and cortical bone. Spongy bone is also called cancellous bone, trabecular bone and medullary bone. [Another term, lamellar bone, is sometimes used to mean only compact bone, and sometime used to denote mature bone, whether spongy or compact, as opposed to immature bone. Therefore, the term lamellar bone is a bit ambiguous].

Compact bone and spongy bone are found in specific locations. In long bones, most of the thickness of the diaphysis is made of compact bone, with a small amount of spongy bone facing the marrow cavity. The ends (epiphyses) of long bones, however, consist mostly of spongy bone covered with a shell of compact bone. The flat bones of the skull have a middle layer of spongy bone sandwiched between two relatively thick layes of compact bone.

Compact Bone: Compact bone is composed of cylindrical structures called osteons or Haversian systems. An osteon consists of concentric lamellae of bone matrix, mainly collagen fibres, surrounding a central canal called the Haversian canal, which contains small blood vessels and nerves. The long axis of osteons is usually parallel to the long axis of the bone. The collagen fibres within any one lamella are generally parallel with one another, but the collagen fibres in the different lamellae of an osteon are oriented at different angles. This increases the strength of the osteon. Canals called Volkmann's canals link the Haversian canals of different osteons with one another and with the marrow cavity. They provide the major route for blood vessels from the marrow cavity to the Haversian canals of osteons. Between the lamellae of an osteon are lacunae containing bone cells called osteocytes. Canaliculi connect the lacunae with one another and with the Haversian canal. The canaliculi contain the processes of the osteocytes, which communicate with one another via gap junctions. Thus nutrients and other substances, such as hormones, can pass from blood vessels in the Haversian canal to distant osteocytes via a sort of bucket brigade. Osteocytes can be involved in both bone deposition and bone resorption.

In addition to the lamellae of osteons, lamellae not belonging to any osteon can be seen in compact bone. These are called interstitial lamellae and are the remnants of previous osteons. They reflect the fact that bone is not static but is constantly being remodelled. In addition, the inner and outer surfaces of long bone have lamellae that run the length of the shaft. They are called, respectively, the inner and outer circumferential lamellae.

Spongy Bone: Spongy bone is composed of bone spicules, also called trabeculae, of varying shapes and sizes. The spaces between the spicules are filled with marrow. The composition of spongy bone (cells and matrix) is the same as that of compact bone. In spongy bone, however, the lamellae of collagen are not arranged concentrically around a central canal, but run parallel to one another. Osteocytes sit in lacunae between lamellae.

Sharpey's Fibres

Collagen fibres from tendons and ligaments extend into bone tissue at an angle, and become continuous with the collagen fibres of the bone matrix.

Intramembranous Ossification, Endochondral Ossification

In the embryo, bone tissue arises through two processes, intramembranous ossification and endochondral ossification. In intramembranous ossification, bone is formed directly from mesenchymal tissue. The flat bones of the skull and face, the mandible and the clavicle develop in this manner. In endochondral ossification, a cartilage model of the bone is formed first, and is later replaced by bone. The weight-bearing bones of the axial skeleton and the bones of the extremities (= most of the skeleton) develop in this manner.

The first bone to arise, whether from mesenchyme or from cartilage (or in fracture repair postnatally), is in the form of spicules. These first spicules are made of immature bone, also called woven bone. In immature bone, the collagenous lamellae are not arranged in parallel or concentric arrays (as in mature spongy and compact bone, respectively), but are randomly oriented and loosely intertwined (hence woven). Immature bone also has more ground substance than mature bone. Consequently, immature and mature bone show different staining characteristics, immature bone stains more with hematoxylin and mature bone more with eosin. The spicules of immature bone are remodelled. The remodelling process can eventually give rise to more spongy bone or to compact bone. The remodelling of bone continues throughout life (remember those interstitial lamellae in compact bone represent former osteons). Immature bone is the predominant bone in the developing fetus. In the adult, most immature bone is replaced by mature bone, but immature bone is seen where bone is being remodelled or repaired, and in certain specific areas, such as the alveolar sockets of the oral cavity. Because most of the spongy bone (= in form of spicules) you look at in the lab is from the embryo, you may erroneously come to think of spongy bone as synonymous with immature bone.

When bone matrix is first secreted, it is not yet mineralized and is called osteoid. As mentioned above, osteoblasts also bring about the mineralization of bone.

Intramembranous Ossification: The first step in intramembranous ossification is the aggregation of mesenchyme cells in the area where bone is to be formed. The tissue in this area becomes more vascularized, and the mesenchyme cells begin to differentiate into osteoblasts, which secrete the collagen and ground substance (proteoglycans) of bone matrix (collectively called osteoid). The osteoblasts maintain contact with one another via cell process. The osteoid becomes calcified with time, and the processes of the cells (called osteocytes when they are surrounded with matrix) become enclosed in canaliculi. Some of the mesenchymal cells surrounding the developing bone spicules proliferate and differentiate into osteoprogenitor cells. Osteoprogenitor cells in contact with the bone spicule become osteoblasts, and secrete matrix, resulting in appositional growth of the spicule. Intramembranous ossification begins at about the eighth week in the human embryo.

Endochondral Ossification: Endochondral ossification also begins with the aggregation of mesemchyme cells, but these differentiate into chondroblasts which secrete hyaline cartilage matrix. The cartilage is secreted in the general shape of the bone that it will become, and grows by both interstitial (mostly in length) and appositional (mostly in width) growth.

Sometime during the growth of this cartilage model (starting at about week 12 in the human fetus), some of the inner perichondrial cells begin to give rise to osteoblasts instead of chondroblasts. (As a result, the former perichondrium is now called the periosteum.) In long bones, this process begins at the mid-region of the bone. The newly formed osteoblasts secrete osteoid, forming a bone collar around the cartilage model. Therefore the very first bone that is formed during endochondral ossification is considered to arise by intramembranous ossification.

With the formation of the bone collar, the cartilage cells in the underlying cartilage begin to hypertrophy and secrete alkaline phosphatase. The surrounding cartilage matrix becomes calcified. The bone collar and the calcified matrix inhibit diffusion of nutrients, and the chondrocytes begin to die. The matrix breaks down, and the lacunae of the dying chondrocytes become confluent making a large cavity. At the same time, blood vessels grow through the bone collar, bringing osteoprogenitor (periosteal) cells with them. These differentiate into osteoblasts when they come in contact with the calcifying cartilage matrix. Osteoid is deposited on the calcifying matrix which is eventually removed. (This is the endochondral part of endochondral ossification.) The bone spicules are remodelled by osteoclasts and osteoblasts. The invading blood vessels also bring with them cells which dfferentiate into the hematopoietic cells of the bone marrrow. In long bones, the midregion where bone formation is initiated is called the primary ossification centre. Secondary ossification centres will develop after birth at the ends (epiphyses) of the bone. During the years of growth, the primary and secondary ossification centres are separated by a cartilage plate called an epiphyseal (growth) plate. This plate allows the bone to grow in length.

Epithelia

Cells that line surfaces: Epithelium is the name given to the cells that line a surface. The outer surface of the body is lined by an epithelium called the epidermis. The inside of the body cavity is also lined by an epithelium, as is the outer surface of many internal organs. The inner surfaces of all the tubes and ducts of the body are also lined by epithelia, whether they open to the outside (eg. digestive system, reproductive system) or are enclosed (eg. cardiovascular system). Glands (eg. sweat glands, salivary glands, liver) are also formed by epithelial cells.

Barrier and other functions: A major function of all epithelia is to form a barrier. Thus, epithelial cells are always contiguous with one another and are usually joined by special junctions. A discussion of the types of junctions is found in Ross et al. Generally, the intercellular space between epithelial cells is very limited. The functions of epithelia differ markedly. Although all form a barrier, some are much more impermeable than others. The epidermis of the skin is a more or less impermeable barrier. On the other hand, the epithelium lining the intestine, while providing a barrier, must also absorb nutrients. Some epithelia have a secretory function (eg. stomach, glands), and some are actively involved in the synthesis of a large number of metabolic substances (eg. the endothelial cells lining blood vessels).

Free surface and basement membrane: All epithelia have a free surface, that is to say a surface to which no cellular or extracellular formed elements adhere. The free surface may contain specialized modifications, such as microvilli or cilia. Such modifications enhance the function of the epithelium. Microvilli in the intestine greatly increase the absorptive surface, while cilia in the respiratory tract beat mucous secretions and entrapped particles away from the lungs toward the pharynx. A discussion of cell surface modifications is found in Ross et al. All epithelia rest on a basement membrane, through which it is attached to the underlying tissue. The basement membrane is generally not identifiable with the light microscope using standard hematoxylin and eosin (H&E) staining. That is because it is thin and also stains with eosin, which makes it blend in with the connective tissue immediately adjacent to it. When a "basement membrane" appears unusually thick, as in the trachea, it is due to a dense layer of closely spaced and aligned collagen fibrils belonging to the underlying connective tissue. The basement membrane can be identified in light microscopy with periodic acid-Schiff (PAS) staining, which reacts with the sugar moieties of its proteoglycans.

With the electron microscope, the basement membrane can be seen to consist of several components. An electron dense layer, called the basal lamina (or lamina densa) consists of a network of fine filaments. Other filaments in a layer called the lamina lucida join the basal lamina to the plasma membrane of the epithelial cells, while anchoring fibrils or microfibrils connect it to reticular fibres of the underlying connective tissue. For a more detailed discussion of the basement membrane, see Ross et al. Although the term basement membrane was originally used with light microscopy, and the term basal lamina referred only to the electron dense layer seen with the electron microscope, the two terms are now used interchangeably by many people.

Types of epithelium: Epithelia are classified according to the number of layers of cells they contain and the shape of the cells. If the cells are arranged in a single layer, the epithelium is called simple; if the cells are arranged in two or more layers, the epithelium is called stratified. If the cells have a flattened shape, the epithelium is called squamous; if the cells have about the same height, width and depth, it is called cuboidal; if the height of the cells exceeds their width, it is called columnar. Thus, an epithelium with a single layer of flattened cells is called a simple squamous epithelium. If the shape of the cells is not identical in the various layers of a stratified epithelium, it is the outermost layer (i.e. at the free surface) that determines the classification. For example, the epidermis is called a stratified squamous epithelium even though the cells in the basal layers are cuboidal, because the outermost layer of cells is squamous. Images of different types of epithelia are shown below.

Simple Squamous Epithelium

A simple squamous epithelium consists of a single layer of flattened cells. The arrows indicate the basophilic nuclei of the epithelial cells, the eosinophilic cytoplasm is difficult to distinguish from the surrounding kidney tubules.

Connective Tissue

A Really Diverse Group of Tissues

Connective tissue is the most diverse of the four tissue types with a wide variety of functions. It ranges in consistency from the gel-like softness of areolar connective tissue to the hardness of bone. Blood is also a connective tissue. Connective tissue (CT) forms an extensive compartment in the body and can be considered as the "glue" that holds the body together. A number of the specialized connective tissues.

Connective Tissues have Cells and an Extracellular Matrix

Connective tissue consists of cells and extracellular material secreted by some of those cells. Thus, unlike the other basic tissues (epithelia, muscle, nervous), the cells in CT may be widely separated from one another within the extracellular matrix. In many types of connective tissue, the matrix-secreting cells are called fibroblasts. Frequently, an abundance of other cell types (eg. macrophages, mast cells, lymphoid cells) may also be present.

The Extracellular Matrix Consists of Ground Substance and Fibres

Ground substance: The ground substance occupies the space between the cells and fibres of connective tissues. It has a high water content, and in specially prepared sections, has an amorphous appearance. In routine preparations, it is lost during the fixation and dehydration process, and only cells and fibres can be seen. Thus, in typical histological sections, the functional importance of ground substance is not at all evident.

Ground substance consists largely of proteoglycans and hyaluronic acid. Proteoglycans are very large macromolecules, consisting of a core protein to which many glycosaminoglycan (GAG) molecules are covalently attached, somewhat like the bristles around the stem of a bottle brush. GAGs, (which used to be called mucopolysaccharides, a name you might still occasionally encounter), are long-chained polysaccharides made up of repeating disaccharide units. One of the sugars in each disaccharide unit is a hexosamine (also called glycosamine), hence the name GAG. Many of the sugars in GAGs have sulfate and carboxyl groups, which makes them highly negatively charged. The high density of negative charges attracts water, forming a hydrated gel. This gel permits the rapid diffusion of water-soluble molecules but inhibits the movement of large molecules and bacteria. A family of seven distinct GAGs is recognized, based on differences in the specific sugar residues, the nature of the linkages and the degree of sulfation. Hyaluronic acid (HA) is itself a GAG, but it differs in several respects from typical GAGs. It is extremely long and rigid, consisting of a chain of several thousand sugars, as opposed to several hundred or less in other GAGs. It also does not bind to a core protein to become part of a proteoglycan. Instead, proteoglycans indirectly bind to HA via special linker proteins, to form giant macromolecules. These hydrophilic macromolecules are particularly abundant in cartilage ground substance, and are responsible for the turgor pressure that gives cartilage its shape.

Fibres: There are three types of fibres secreted by connective tissue cells: collagen fibres, reticular fibres, and elastic fibres. The abundance and preponderance of different types of fibres varies in different CTs. Each type of fibre is formed by proteins made of long peptide chains.

Collagen fibres: The most common fibre type is the collagen fibre. These are flexible fibres with a high tensile strength. In typical preparations for the light microscope (LM), they appear as wavy lines of variable width and indeterminate length.

With the electron microscope, collagen fibres are seen to be made up of thread-like subunits called collagen fibrils. Each fibril, in turn, is made up of collagen molecules that are aligned, head to tail, in overlapping rows. Within each row, there is a gap beween the tail of one molecule and the head of the next. The fibril's strength is due to covalent bonds between collagen molecules of adjacent rows-not the head to tail attachment within a row.

The collagen molecule (called tropocollagen) is composed of three intertwined polypeptide chains (each of which is called an alpha chain) that form a right-handed triple helix. Except for the ends of the chain, every third amino acid is a glycine. A hydroxyproline frequently precedes each glycine, and a proline frequently follows each glycine. Sugar groups are associated with the triple helix, so collagen is properly called a glycoprotein.

The alpha chains that form the helix are not all alike, and, based on differences within the chains, as many as 16 types of collagen have been identified. They are classified by Roman numerals on the basis of chronology of discovery. Type I collagen is the most prevalent type of collagen, and constitutes about 90% of body collagen. It is the collagen found in the dermis of the skin, bone, tendon, organ capsules and many other areas. The fibres found in cartilage are finer, they consist of type II collagen. Type IV collagen is found in the basal lamina (of basement membrane) of epithelia. There is no need to memorize where the different types of collagen are found.

Reticular fibres: Reticular fibres are closely related to collagen fibres. They are made of type III collagen fibrils (sometimes in association with type IV collagen). The individual fibrils that constitute the reticular fibre are of narrow diameter and typically do not bundle to form thick fibres. Reticular fibres cannot be identified in routine preparations. They can be displayed with special silver preparations or with the periodic acid-Schiff (PAS) reaction because of their relatively high sugar content.. Reticular fibres were given their name because they are arranged in a mesh-like pattern. They provide a supporting framework for the cellular constituents of various tissues and organs, for example the liver. They are also found at the boundary of the CT and epithelium in loose CT, around adipocytes, small blood vessels, nerves and muscle cells. In most locations, reticular fibres are produced by fibroblasts. However, the reticular fibres that support the stroma of hemopoietic and lymphatic tissue are made by special cells called reticular cells. Each reticular cell maintains a unique relationship to its fibre, surrounding it with its cytoplasm and thereby isolating it from its environment.

Other areas where reticular fibres are not produced by fibroblasts include the endoneurium of peripheral nerves, where they are produced by Schwann cells (discussed under Nervous Tissue below), the tunica media of blood vessels and the muscularis externa of the alimentary canal. The tunica media and muscularis externa are made up of smooth muscle cells, and it is the smooth muscle cells themselves that secrete all the CT fibre types within those structures.

Elastic fibres: Elastic fibres are thinner than collagen fibres and are arranged in a branching pattern to form a three dimensional network. They give tissue the ability to cope with stretch and distension. Elastic fibres are interwoven with collagen fibres in order to limit distensibility and to prevent tearing. Elastic fibres are composed of two structural components: elastin and microfibrils. Elastin is a protein related to collagen but with an unusual polypeptide backbone that causes it to coil in a random way. The configuration of one molecule's coiling is not permanent, it oscillates from one shape to another. The coiled elastin molecule can be stretched. When the force causing the stretch is withdrawn, the molecule recoils back to its former state. Two large amino acids unique to elastin, called desmosine and isodesmosine, cause elastin molecules to covalently bond to one another and form an elastin matrix. The entire matrix is engaged during the stretch and recoil of elastic tissue. Microfibrils consist of a fibrillar glycoprotein. In developing elastic tissue, they appear before the elastin, and are believed to serve as an organizing structure for it. (Note: Don't confuse microfibrils, an extracellular structure, with microfilaments, an intracellular structure made of actin.)

Elastic material is found in certain ligaments (elastic ligaments), some cartilage (called elastic cartilage) and in large arteries (elastic arteries). In most cases, the elastic fibres are produced by fibroblasts. In the case of elastic arteries, it is produced by the smooth muscle cells of the tunica media.

However, the elastic material produced by the smooth muscle cells does not contain microfibrils, only elastin, and as a result does not form elastic fibres. Instead, the elastin is laid down in fenestrated (having gaps or openings) sheets or lamellae arranged in concentric layers between layers of smooth muscle. Elastic fibres do not stain very well with eosin and in routine preparations usually cannot be distinguished from collagen fibres. (Certain fixatives cause them to become somewhat refractile, and when this occurs, they can be distinguished from collagen fibres even with H&E staining.) Elastic fibres are selectively stained with special dyes such as orcein and resorcin-fuchsin.

The cells of connective tissues: As mentioned above, many different kinds of cells can be found in CTs. In some CTs, there is a large diversity of cell types, in others, the diversity is very low. Some of the cells in CTs are fixed, that is, they are permanent residents in the connective tissue. Other cells are wandering, they are transient migrants who have entered the CT from the blood in response to specific stimuli. The list below summarizes some of the cells commonly found in connective tissues.

Fixed Cells

Fibroblasts: Fibroblasts are the principal cells of connective tissue. They are responsible for the secretion of all types of fibres (collagen, reticular, elastin) and the complex carbohydrates of ground substance. A single fibroblast is believed to be able to secrete all the extracellular components, both sequentially and simultaneously. In routine histological preparations, only the nucleus of the fibroblast can be identified, the cytoplasmic processes blend in with the surrounding collagen. The nucleus appears as an elongated or discoid structure. Like the nucleus of any cell, fibroblast nuclei stain blue with H&E, as nucleic acids avidly bind to basic dyes.

Chondroblasts & chondrocytes: These are the matrix-secreting cells of cartilage.

Osteoblasts & osteocytes: These are the matrix-secreting cells of bone.

Macrophages: Macrophages, also called histiocytes, are phagocytic cells derived from monocytes.

Adipose cells: Also called adipocytes, these cells are specialized to store neutral fat.

Mast cells: Mast cells have granules containing histamine, heparin and anaphylactic factors. When released in response to an antigen, they cause hypersensitivity reactions, allergy and anaphylaxis.

Undifferentiated mesenchyme cells: These are cells that retain the multiple potentials of embryonic mesenchyme cells. They are found in the tunica adventitia (the outer layer of CT) of venules.

Wandering Cells

Lymphocytes: These are cells responsible for immune responses that circulate in the blood. Normally, only small numbers are found in the CTs throughout the body. The number increases dramatically at certain sites of tissue inflammation.

They are also very numerous in the lamina propria of the respiratory and gastrointestinal tracts, where they are involved in immunosurveillance. The lamina propria is a layer of loose CT lying immediately beneath the epithelium.

Plasma cells: Plasma cells are derived from B-lymphocytes and produce antibodies against a specific antigen. They have a limited migratory ability and a short life.

Neutrophils: Neutrophils are white blood cells that act as phagocytes in the early stages of acute inflammation.

Eosinophils: Eosinophils are white blood cells that are found in the lamina propria of the GI tract, and at sites of allergic reaction and parasitic infection.

Basophils: Basophils are white blood cells that are similar to mast cells in having vasoactive agents released in response to an allergen.

Monocytes: Monocytes are white blood cells that will give rise to all the phagocytes of the mononuclear phagocytic system.

Classification of Connective Tissues

Connective tissues are classified on the basis of types and relative abundance of cells, fibres and ground substance, and on the organization of fibres. The images below will introduce you to the diversity of connective tissues. You will study a number of these connective tissues in greater detail in some of the Blocks.

Loose (or Areolar) Connective Tissue

This is a cellular type of connective tissue, with abundant ground substance and thin and relatively sparse fibres. It has a viscous gel-like consistency and is important for the diffusion of oxygen and nutrients from small vessels, and the diffusion of metabolites back to the vessels. The primary location of loose connective tissue is beneath epithelia that line the internal surfaces of the body, in association with the epithelia of glands and around small vessels. It is the initial site at which antigens, bacteria and other agents that have breached an epithelial surface can be destroyed.

Dense Irregular Connective Tissue

In dense irregular CT, collagenous fibres make up the bulk of the tissue. Fibroblasts are scarce and usually the only cell type present. Little ground substance is present. The abundance of collagenous fibres gives the tissue strength. The fibres are typically arranged in bundles in various directions (hence irregular), which enables the tissue to withstand various stresses. Dense irregular CT is found on the outside of many organs, in the dermis of the skin and as a distinct layer, called the submucosa, within various organs.

Dense Regular Connective Tissue

In dense regular CT, collagenous fibres are packed in dense regular arrays, between which lie rows of cells. Dense regular CT is found in tendons (which connect muscles to bones), ligaments (which connect bones to bones, some also contain large amounts of elastic fibres and are called elastic ligaments), and aponeuroses (broad flattened tendons whose fibres are arranged in multiple layers, within each layer the fibres are regularly arranged).

Adipose Tissue

Adipocytes, which are specialized to store fat, are found throughout loose connective tissue. When adipocytes are the predominant cell present, the tissue is called adipose tissue. In white or unilocular fat, adipocytes contain a single, large lipid inclusion surrounded by a thin layer of cytoplasm. The lipid mass compresses the nucleus to an eccentric position, producing a "signet ring" appearance.

Muscle Tissue

If it contracts, it's muscle: Muscle tissue is categorized on the basis of a functional property: the ability of its cells to contract. In muscle tissue, the bulk of the cytoplasmic volume consists of the contractile protein fibrils actin and myosin. Muscle is responsible for movement of the body and changes in the size and shape of internal organs. Muscle cells are generally referred to as muscle fibres. (Note that the term "fibre" is used both for muscle cells, and for the extracellular elements, eg. collagen, produced by connective tissue cells.) Muscle fibres are typically arranged in parallel arrays, allowing them to work together effectively.

The structure and physiology of muscle will be studied in detail in the Homeostasis and Development Block. Here, we will discuss only the identifying features of the different types of muscle. Additional information on the histology of muscle is found in Ross et al.

The three types of muscle: Three types of muscle tissue can be identified histologically: skeletal muscle, cardiac muscle and smooth muscle. The fibres of skeletal muscle and cardiac muscle exhibit cross striations at the light microscope level and they are both referred to as striated muscle.

Skeletal Muscle

Skeletal muscle constitutes the muscle that is attached to the skeleton and controls motor movements and posture. There are a few instances where this type of muscle is restricted to soft tissues: the tongue, pharynx, diaphragm and upper part of the esophagus. (Some people use the term visceral striated muscle in the foregoing examples, but since it is identical in structure to the muscle that moves the skeleton, we won't bother with the extra term.)

Skeletal muscle fibres (cells) are actually a multinucleated syncytium formed by the fusion of individual small muscle cells or myoblasts, during development. They are filled with longitudinally arrayed subunits called myofibrils. The myofibrils are made up of the myofilaments myosin (thick filaments) and actin (thin filaments). The striations reflect the arrangement of actin and myosin filaments and support structures. The individual contractile units are called sarcomeres. A myofibril consists of many sarcomeres arranged end to end. The entire muscle exhibits cross-striations because sarcomeres in adjacent myofibrils and muscle fibers are in register. The most obvious feature in longitudinal sections of skeletal muscle is the alternating pattern of dark and light bands, called respectively the A (anisotropic) and I (isotropic) band. The I band is bisected by a dense zone called the Z line, to which the thin filaments of the I band are attached.

The nuclei are located peripherally, immediately under the plasma membrane (sarcolemma). The thickness of individual muscle fibres varies (depending for example on location in the body and exercise) but each fibre is of uniform thickness throughout its length. Skeletal muscle fibres do not branch. Connective tissue elements surround muscle fibres. Individual muscle fibres are surrounded by a delicate layer of reticular fibres called the endomysium. Groups of fibres are bundled into fascicles by a thicker CT layer called the perimysium. The collection of fascicles that constitutes one muscle is surrounded by a sheath of dense CT called the epimysium, which continues into the tendon. Blood vessels and nerves are found in the CT associated with muscle. The endomysium contains only capillaries and the finest neuronal branches.

Summary: Skeletal muscle fibres bear obvious striations, have many peripherally located nuclei, are of the same thickness throughout their length and do not branch.

Cardiac Muscle

Cardiac muscle is the type of muscle found in the heart, and at the base of the venae cavae as they enter into the heart. Cardiac muscle is intrinsically contractile but is regulated by autonomic and hormonal stimuli. Cardiac muscle exhibits striations because it also has actin and myosin filaments arranged into sarcomeres. Generally these striations do not appear as well-defined as in skeletal muscle. (At the ultrastructural level, some differences in the arrangement of the sarcoplasmic retiuculum and T tubules can be seen. Cardiac muscle also has a much greater number of mitochondria in its cytoplasm. More details on the anatomy and physiology of muscle will be discussed in H&D and Cardiovascular Blocks.)

At the light microscope level, a number of features distinguish cardiac from skeletal muscle. Cardiac muscle cells have only one or two nuclei, which are centrally located. The myofibrils separate to pass around the nucleus, leaving a perinuclear clear area (not always evident in standard preparations). This clear area is occupied by organelles, especially mitochondria (which are of course not visible in

LM). As in skeletal muscle, individual muscle fibres are surrounded by delicate connective tissue. Numerous capillaries are found in the connective tissue around cardiac muscle fibres. Cardiac muscle cells are joined to one another in a linear array. The boundary between two cells abutting one another is called an intercalated disc. Intercalated discs consist of several types of cells junctions whose purpose is to facilitate the passage of an electrical impulse from cell to cell and to keep the cells bound together during constant contractile activity. Unlike skeletal muscle fibres, cardiac muscle fibres branch and anastomose with one another. Although made up of individual fibres, heart muscle acts as a functional syncytium during contraction for the efficient pumping of blood.

Specialized fibres, called Purkinje fibres, arise from the atrioventricular node and travel along the interventricular septum toward the apex of the heart, sending branches into the ventricular tissue. Purkinje fibres are of larger diameter than ordinary cardiac fibres, with fewer myofibrils and an extensive, well-defined clear area around the nucleus. They conduct impulses at a rate about four times faster than that of ordinary cardiac fibres and serve to coordinate the contraction of the atria and ventricles.

Summary: Cardiac muscle fibres are striated, have one or two centrally located nuclei, branch and anastomose with other fibres, and are joined to one another by intercalated discs.

Smooth Muscle

Smooth muscle is the intrinsic muscle of the internal organs and blood vessels. It is also found in the iris and ciliary body of the eye and associated with hair follicles (arrector pili). No striations are present in smooth muscle due to the different arrangement of actin and myosin filaments. Like cardiac muscle, smooth muscle fibres are intrinsically contractile but responsive to autonomic and hormonal stimuli. They are specialized for slow, prolonged contraction.

Smooth muscle fibres are generally arranged in bundles` or sheets. Each fibre is fusiform in shape with a thicker central portion and tapered at both ends. The single nucleus is located in the central part of the fibre. Fibres do not branch. They range enormously in size, from 20 (in wall of small blood vessels) to 500 (in wall of uterus during pregnancy) micrometers. Smooth muscle fibres lie over one another in a staggered fashion (tapered part of one fibre over thicker part of another). In longitudinal sections, it is often not possible to distinguish the fibre boundaries, and smooth muscle may closely resemble connective tissue (bundles of collagen). Where smooth muscle bundles are interlaced with bundles of connective tissue (eg. in the uterus), one can distinguish the smooth muscle by the orientation of the nuclei (all oriented in the same direction), and the greater abundance of nuclei per unit area (every smooth muscle cell has a nucleus, fibroblast nuclei are more scattered in bundles of CT). Also, smooth muscle nuclei often have a corkscrew shape in longitudinal section due to contracton of the muscle fibre during fixation. In cross section, smooth muscle appears as profiles of various sizes, depending on whether the cut went through the thick central part or tapered end of any individual fibre. Nuclei are seen only in the thicker profiles.

One distinguishing physiological feature of smooth muscle is its ability to secrete connective tissue matrix. In the walls of blood vessels and the uterus in particular, smooth muscle fibres secrete large amounts of collagen and elastin.

Summary: Smooth muscle fibres are fusiform with tapered ends, have a single centrally located nucleus, and do not exhibit striations.

Nervous Tissue

Nervous tissue is designed to transmit electrical impulses: Nervous tissue consists of nerve cells and associated supporting cells. All cells exhibit electrical properties, but nerve cells, also called neurons, are specifically designed to transmit electrical impulses from one site in the body to another, and to receive and process information. Supporting cells are non-conducting cells that are in intimate physical contact with neurons. They provide physical support, electrical insulation and metabolic exchange with the vascular system.

What makes it a nerve cell?: Nerve cells are very variable in appearance, but all neurons have a cell body, also called soma or perikarion, and processes. The cell body contains the nucleus and organelles that maintain the nerve cell. The processes extend from the nerve cell to communicate with other cells. There are two types of processes: dendrites that receive impulses and axons that transmit impulses. All nerve cells have an axon (usually only one), which is usually the longest process that extends from the cell. Most nerve cells have dendrites, usually many, and these are generally shorter and thicker than the axon. Nerve cells can also receive impulses right on the nerve cell body, and some nerve cells have no dendrites. The junction where a nerve cell communicates with another nerve cell or an effector cell (eg. muscle fibre) is called a synapse. The terminal part of the axon releases a chemical called a neurotransmitter which acts on the membrane of the other cell. Many axons are wrapped in a lipid-rich covering called myelin. This myelin sheath insulates the axon from the surrounding extracellular component and increases the rate of electrical conduction. The myelin sheath is discontinuous at intervals called the nodes of Ranvier. The areas covered with myelin are called internodal areas. In myelinated axons, the voltage reversal (that is, the impulse propagation) can occur only at the nodes, and the impulse "jumps" from node to node. This is called saltatory conduction. In unmyelinated axons, the impulse is conducted more slowly, moving as a wave of voltage reversal along the axon.

Brief overview of the organization of the nervous system: The nervous system is divided into the central nervous system (CNS) and the peripheral nervous system (PNS). The central nervous system is limited to the brain and spinal cord, all other nervous tissue belongs to the peripheral nervous system. An individual nerve cell can have components in both the CNS and the PNS. For example, the cell body of a motor neuron (one that causes skeletal muscle to move) lies in the CNS, but its axon leaves the CNS to travel in a peripheral nerve to reach the muscle it innervates. In many instances nerve cell bodies are found together in clusters. Clusters of nerve cell bodies in the CNS are called nuclei (sing. nucleus, not to be confused with the nucleus of every cell that houses its DNA etc.). Clusters of nerve cell bodies in the peripheral nervous system are called ganglia (sing. ganglion).

In the central nervous system, areas containing nerve cell bodies, their myelinated and unmyelinated processes and supporting (glial) cells are called grey matter. Areas containing predominantly myelinated axons but also some unmyelinated axons and glial cells are referred to as white matter, because in fresh preparations the myelinated tissue looks white. In the brain, the grey matter is exterior to the white matter, the reverse is the case in the spinal cord.

In the CNS, there is a selective restriction of blood-borne substances called the blood-brain barrier. This barrier is formed by special tight junctions in the endothelial cells of the blood vessels. Certain specific parts of the brain are "leakier" than the rest-otherwise hormones couldn't act on the brain to influence behaviour. The peripheral nervous system has motor components called efferent neurons (because they arise in the CNS and head away from it to the periphery), and sensory components called afferent neurons (because the impulse is initiated in a specialized receptor in the PNS and heads toward the CNS). A special component of the peripheral nervous system is the autonomic nervous system (ANS). This is the portion of the PNS that conducts impulses to smooth muscle, cardiac muscle and glandular epithelium. It is classified into three divisions: the sympathetic, parasympathetic and enteric divisions.

You will be learning more about the structure of nerve cells and the organization of the nervous system in the Homeostasis and Development Block and in the Neurology Block.

Support cells play a vital role: Support cells are essential to the function and survival of nerve cells. The CNS and PNS each have their own specific types of support cells.

Support cells in the CNS: The general term for support cells in the CNS is glia or neuroglia (a.k.a. glial cells, neuroglial cells). There are three types of neuroglial cells. (1) Oligodendrocytes, the myelin-secreting cells of the CNS. (2) Astrocytes, which provide physical and metabolic support for nerve cells. (3) Microglia, or microglial cells (a.k.a. mesoglia), which are the phagocytes of the CNS.

Oligodendrocytes: As their name implies, oligodendrocytes have few processes. They are often found in rows between axons. The myelin sheath around axons is formed by concentric layers of oligodendrocytes plasma membrane. Each oligodendrocyte gives off several tongue-like processes that find their way to the axon, where each process wraps itself around a portion of the axon, forming an internodal segment of myelin. Each process appears to spiral around its segment of the axon in a centripetal manner, with the continued insinuation of the leading edge between the inner surface of newly formed myelin and the axon. One oligodendrocyte may myelinate one axon or several. The nucleus-containing region may be at some distance from the axon(s) it is myelinating. In the CNS, nodes of Ranvier (between myelinated regions) are larger than those of the PNS, and the larger amount of exposed axolemma makes saltatory conduction more efficient.

Unmyelinated axons in the CNS are truly bare, that is they are not embedded in any glial cell process. (In contrast to the situation in the PNS, described below.)

Bibliography

Archer, M., S. J. Hand, and H. Godthelp: *Riversleigh. The Story of Animals Reproduction in Ancient Rainforests*, Reed Books, Balgowhah, 1991.

Arseniev, V.A.: *Atlas of Marine Mammals*, Neptune City: T. F. H. Publishing, Inc., 1986.

Billingham, R.E., and W.K. Silvers. : *Transplantation of Tissues and Cells*, Wistar Inst. Press, Philadelphia, 1961.

Bower, B.: *Fossils may Clarify Mammal Evolution*, Science News, 1984.

Bronson, F. H.: *Mammalian Reproductive Biology*, Univ. Chicago Pr., Chicago, 1990.

Clutton-Brock, J. : *A Natural History of Domesticated Mammals*, Cambridge Univ. Pr., New York, 1999.

Cooper, G., and A. L. Schiller: *Anatomy of the Guinea Pig*, Harvard Univ. Pr., Cambridge, 1975.

Daniel, J.C. Jr. : *Methods in Mammalian Reproduction*, Orlando, FL, Academic Press, 1978.

Eisenberg, John E.: *The Mammalian Radiations*, The University of Chicago Press, 1981.

Flowerdew, J. R. : *Animals: Their Reproductive Biology and Population Ecology*, Cambridge Univ. Pr., New York, 1987.

Gordon, G. A. : *Animals Physiology,* Harper and Row, New York, 1989.

Gray, J.: *Animal Locomotion*, Norton, New York. 1968.

Greer, A. E.: *Notes on the Mode of Reproduction in Anguid Lizards*, Herpetologica, 1967.

Griffin, D.R.: *Animal Minds*, University of Chicago Press. Chicago, 1992.

Grzimek, B.: *Grzimek's Animal Life Encyclopedia*, McGraw Hill, New York, 1989.

Harrison, R. J.: *Functional Anatomy of Marine Animals*, New York: Academic Press, 1974.

Hill, R. W., and G. A. Gordon. : *Animal Physiology*, Harper and Row, New York, 1989.

Jones, M.B. & Lazenby, A. : *The Grass Crop*, London: Chapman & Hall, 1988.

Martin, A.M. : *Fisheries Processing : Biotechnological Applications*, Chapman and Hall, Delhi, 2009.

Matthews, L. H.: *The Life of Animals Reproduction*, London, Weidenfield and Nicholson, 1969.

Raymond, F., Redman, P., & Waltham, R. : *Forage conservation and Feeding*. Ipswich: Farming Press, 1986.

Skerman, P.J., & Riveros, F. : *Tropical Grasses*. Rome: FAO, 1989.

Skerman, P.J., Cameron, D.G., & Riveros, F. : *Tropical Forage Legumes*, Rome: FAO, 1988 .

Sweeney, Del. : *Agriculture in the Middle Ages: Technology,Practice, and Representation*, Philadelphia: University of Pennsylvania Press, 1995.

Whyte, R.O, Moir, T.R.G., & Cooper, J.P. : *Grasses in Agriculture*, Rome: FAO, 1959.

Index

❑❑❑